XXXV Symposium on Nuclear Physics 2012

Journal of Physics: Conference Series Volume 387

Morelos, Mexico
3 – 6 January 2012

Editors:

E. Padilla-Rodal
R. Bijker

ISBN: 978-1-62276-824-0
ISSN: 1742-6588

Printed from e-media with permission by:

Curran Associates, Inc.
57 Morehouse Lane
Red Hook, NY 12571

Some format issues inherent in the e-media version may also appear in this print version.

Printed by Curran Associates, Inc. (2013)

For permission requests, please contact the Institute of Physics
at the address below.

Institute of Physics
Dirac House, Temple Back
Bristol BS1 6BE UK

Phone: 44 1 17 929 7481
Fax: 44 1 17 920 0979

techtracking@iop.org

Additional copies of this publication are available from:

Curran Associates, Inc.
57 Morehouse Lane
Red Hook, NY 12571 USA
Phone: 845-758-0400
Fax: 845-758-2634
Email: curran@proceedings.com
Web: www.proceedings.com

TABLE OF CONTENTS

Outstanding Features Of Reactions With Halo Nuclei 1
E. Aguilera

Interaction Spatial Barrier Distributions For The Proton-Halo System $^{8}B+^{58}Ni$ 9
A. Camacho, E. Aguilera

High Precision Measurements For The Rp-Process 17
G. Berg, Y. Fujita, J. Gorres, M. Harakeh, K. Hatanaka, A. Long, R. Neveling, F. Smit, R. Talwar, M. Wlescher

The FrPNC Experiment, Weak Interaction Studies In Francium At TRIUMF 24
E. Gomez, S. Aubin, R. Collister, J. Behr, G. Gwinner, L. Orozco, M. Pearson, M. Tandecki, D. Sheng, J. Zhang

Systematics of Low Energy Collective States in Neutron-Rich Cd Isotopes 29
J. Batchelder, N. Brewer, C. Gross, R. Grzwacz, J. Hamilton, M. Karny, A. Kuzniak, M. Madurga, A. Mendez, S. Liu, K. Miernik, D. Miller, S. Padgett, S. Paulauskas, K .Rykaczewski, A. Ramayya, D. Stracener, M. Wolinska-Cichocka, E. Zganjar

Superallowed $0^{+}\rightarrow 0^{+}$ Beta-decay from T_z = -1 sd-Shell Nuclei 39
J. Hardy, I. Towner

Nuclear And Weak Interaction Aspects Of Neutrinoless Double Beta Decay: Recent Results 48
O. Civitarese, J. Suhonen

The Neutrino Mixing Angle θ_{13} in an S_3 Flavour Symmetric Model 53
F. Canales, A. Mondragon

Novel Bose-Einstein Interference in the Passage of a Fast Particle in a Dense Medium 63
C. Wong

Conventional vs Gauge Couplings in Elastic πN Scattering 73
A. Mariano, C. Barbero, D. Badagnani

Strangeness of the Proton 79
R. Bijker, J. Ferretti, E. Santopinto

A Panorama Of Excited-State G-Factor Measurements: Advancing Moment Measurements On Radioactive Beams 86
A. Stuchbery

Precise Coulomb Excitation B(E2) Measurements For First 2^{+} States Of Projectile Nuclei Near The Doubly Magic Nuclei ^{78}Ni And ^{132}Sn 95
A. Galindo-Uribarri

Gamma-Ray Spectroscopy At TRIUMF-ISAC: Recent Highlights And Future Plans 103
G. Ball

Density Functional Approaches To Nuclear Dynamics 111
T. Nakatsukasa, S. Ebata, P. Avogadro, L. Guo, T. Inakura, K. Yoshida

Ab Initio No-core Shell Model Calculations in a SU(3)-based Coupling Scheme 121
T. Dytrych, K. Launey, J. Draayer, D. Langr

Symmetry-Adapted Ab Initio Open Core Shell Model Theory 127
J. Draayer, T. Dytrych, K. Launey, D. Langr, A. Dreyfuss, C. Bahri

Systematic Study Of Isoscalar And Isovector Pairing In The 2p1f Shell 132
Y. Lei, S. Pittel, N. Sandulescu, A. Poves, B. Thakur, Y. Zhao

Phase Transitions For Excited States in $^{16}O+\alpha$ ^{20}Ne within the SACM 141
G. Morales-Hernandez, H. Yepez-Martinez, P. Hess

Shell Model Description Of Ge Isotopes 147
J. Hirsch, P. Srivastava

Quantum Phase Transitions In The LMG Model By Means Of Quantum Information Concepts 157
O. Castanos, R. Lopez-Pena, E. Nahmad-Achar, J. Hirsch

Author Index

Journal of Physics: Conference Series **387** (2012) 011001 doi:10.1088/1742-6596/387/1/011001

Preface

The XXXV Symposium on Nuclear Physics was held at Hotel Hacienda Cocoyoc, Morelos, Mexico from January 3th to 6th, 2012.

Conceived in 1978 as a small meeting, over the years and thanks to the effort of various organizing committees, the symposium has become a well known international conference on nuclear physics. To the best of our knowledge, the Mexican Symposium on Nuclear Physics represents the conference series with longest tradition in Latin America and one of the longest-running annual nuclear physics conferences in the world.

The Symposium brings together leading scientists from all around the world, working on the fields of nuclear structure, nuclear reactions, physics with radioactive ion beams, hadronic physics, nuclear astrophysics, neutron physics and relativistic heavy-ion physics. Its main goal is to provide a relaxed environment where the exchange of ideas, discussion of new results and consolidation of scientific collaboration are encouraged.

To celebrate the 35th edition of the symposium we counted with the attendance of 53 colleagues from diverse countries including: Argentina, Australia, Canada, Japan, Saudi Arabia and USA. We were happy to have the active participation of Eli F. Aguilera, Eduardo Andrade, Octavio Castaños, Alfonso Mondragón, Stuart Pittel and Andrés Sandoval who also participated in the first edition of the Symposium back in 1978. We were joined by "old friends" of Cocoyoc (Stuart Pittel, Osvaldo Civitarese, Piet Van Isacker, Jerry Draayer and Alfredo Galindo-Uribarri) as well as several first time visitors that we hope will be back to this scientific meeting in the forthcoming years.

The scientific program consisted of 33 invited talks, proposed by the international advisory committee, which nicely covered the topics of the Symposium giving a balanced perspective between the experimental and the theoretical work that currently is undergoing in each line of research. Fifteen posters complemented the scientific sessions giving the opportunity for Mexican students to present their current research and interact with the visiting scientists.

The present volume contains 21 research articles based on invited talks presented at the symposium. We cannot thank enough to all the authors for their enthusiastic contribution, to the anonymous referees for their time devoted to the review process, which helped us to maintain a high standard of the Conference Proceedings. Finally we would like to thank the International Advisory Committee and the Sponsoring Organizations that made this event possible.

E Padilla-Rodal and R Bijker
Editors

Journal of Physics: Conference Series **387** (2012) 012001 doi:10.1088/1742-6596/387/1/012001

Outstanding features of reactions with halo nuclei

E F Aguilera

Departamento de Aceleradores, Instituto Nacional de Investigaciones Nucleares, Apartado Postal 18-1027, C. P. 11801, México, D.F., México

E-mail: eli.aguilera@inin.gob.mx

Abstract. Recent results for total reaction and fusion excitation functions for halo systems are compiled. The striking similarity in the behavior of total cross sections for proton-halo (^{8}B) and neutron-halo (^{6}He) systems is pointed out. Static and dynamic halo effects are elucidated by comparing data for a neutron-halo nucleus and its core. Fusion cross sections are discussed in terms of a maximum angular momentum for fusion, L_f. A linear energy dependence for L_f is consistent with the data for all the existing measurements so this might actually be a characteristic feature of true halo systems. A qualitative difference between the fusion of neutro-halo and proton-halo systems also is pointed out.

1. Introduction

In the context of nuclear physics research, one of the most relevant events of the last few decades was the advent of physics with radioactive ion beams, a field that is now just over 25 years old. The existence of nuclear halos, which was realized soon after the first experiments, remains as probably the most outstanding discovery in this field. While nuclear halos were actually discovered through reaction measurements at intermediate energies, where static effects are dominant, the respective dynamic effects are expected to become more evident at lower energies, near the corresponding Coulomb barrier. Under these conditions the nucleons in the halo may have time enough to rearrange themselves in the force field of the target so that, for instance, dynamic polarization effects may appear. Up to date, many low energy measurements have been performed with light halo systems, thus giving one the opportunity to search for possible patterns in the respective behavior.

Such a pattern has been found, for instance, in the case of total reaction cross sections, σ_R, for systems involving the neutron-halo projectile ^{6}He. The corresponding excitation functions, when properly scaled to eliminate trivial effects of size and charge, follow a well defined path for all systems studied, which includes a wide range of bombarding energies and target masses. A comparison with respective data for the corresponding core nucleus, ^{4}He, leads to interesting conclusions about the static and dynamic effects of the halo. Although only one excitation function has been measured for the proton-halo projectile ^{8}B, it is quite remarkable that the observed behavior is similar to that of ^{6}He insofar as total reaction cross sections are concerned.

Fusion excitation functions at near barrier energies also have been measured for a few of these systems. Because of the increased importance of direct processes in low energy reactions with halo nuclei, the fusion cross sections are lower than the respective total reaction cross sections. By using a simple single-barrier penetration model, the fusion data can be described in terms of a maximum angular momentum for fusion, L_f. Surprisingly, for all the halo systems analyzed so far, the data seem to indicate a linear energy dependence for L_f.

A discussion of the above topics is the subject of the present work. A compilation of recent results will be presented, showing some of the most prominent halo effects. In the first part, it will be shown that the excitation functions for total reaction cross sections of different systems having as projectiles either the neutron-halo nucleus ^{6}He or the proton-halo isotope ^{8}B, do follow a common systematic trend. In the second part, the existing measurements for fusion of halo systems will be presented and, through the introduction of a simple model to analyze them, a characteristic behavior will be pointed out.

2. σ_R(E) for reactions induced by ^{6}He and ^{8}B

The two-neutron halo nucleus ^{6}He was the first halo projectile for which reaction measurements were done [1], and it is probably the one that has been most extensively studied so far [2]. A recent survey of respective measurements in the low energy regime is given in Ref. [3], where the results of a systematic comparison of total reaction cross sections for systems with different targets are reviewed. However, this kind of comparison was actually motivated originally by the idea of looking for possible signatures of proton-halo effects, as shall be discussed below.

2.1. Comparing reaction cross sections for different systems

Earlier evidence for a possible systematic behavior of total reaction cross sections was observed in Ref. [4], where elastic scattering angular distributions for the proton-halo projectile ^{8}B on a ^{58}Ni target were reported. The measurements were performed at five near barrier energies, using the radioactive ion beam facility TwinSol at the University of Notre Dame [5]. The typical ^{8}B beam in this facility is contaminated with substantial amounts of ^{6}Li and ^{7}Be, so it was possible also to get the respective angular distributions (Fig. 1). The solid curves in the figure correspond to optical model calculations, which in turn give the corresponding total reaction cross sections. So, three total-reaction excitation functions were obtained, one for each system. This was actually the first study of this kind for a proton-halo projectile, which was indeed the main interest of this work, but the other two projectiles are also intrinsically interesting. They are both weakly-bound and ^{7}Be is actually a radioactive nucleus, which in addition happens to be the core for ^{8}B. It was thus interesting trying to compare to each other the results for the three systems. For this purpose, the cross sections and energies were reduced according to the expressions $\sigma_{Reduced} = \sigma_R/(A_p{}^{1/3} + A_t{}^{1/3})^2$, $E_{Reduced} = E_{c.m.}/[Z_pZ_t/(A_p{}^{1/3} + A_t{}^{1/3})]$. This reduction of data is expected to eliminate trivial effects of size and charge without washing out other important effects [6], thus making data for different systems directly comparable to each other.

Intuitively, one would expect this reduction to work better for similar systems, so it was decided to include in the comparison previous data for other halo- and weakly-bound systems having targets of similar masses. The comparison showed very interesting features. In addition to the three systems measured in Ref. [4], data for the weakly-bound projectiles ^{6,7}Li and ^{9}Be were also included, with targets in the neighborhood of ^{58}Ni. As shown in Fig. 2, all data for weakly-bound projectiles fall on the same trajectory, which is indicated with the lower solid curve. As for halo projectiles, in addition to ^{8}B, data for the neutron-halo nucleus ^{6}He were included, the main intention being to spot possible differences between reactions induced by proton-halo vs neutron-halo projectiles. In this case a heavy target (^{209}Bi) was also considered, mainly because the respective data were easily available [7, 8]. In spite of the big difference in

Journal of Physics: Conference Series **387** (2012) 012001 doi:10.1088/1742-6596/387/1/012001

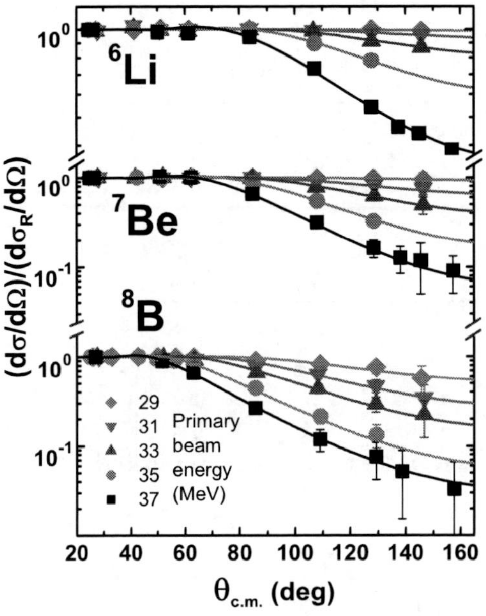

Figure 1. Elastic scattering angular distributions for the (^{8}B,^{7}Be,^{6}Li)+^{58}Ni systems. Taken from Ref. [4].

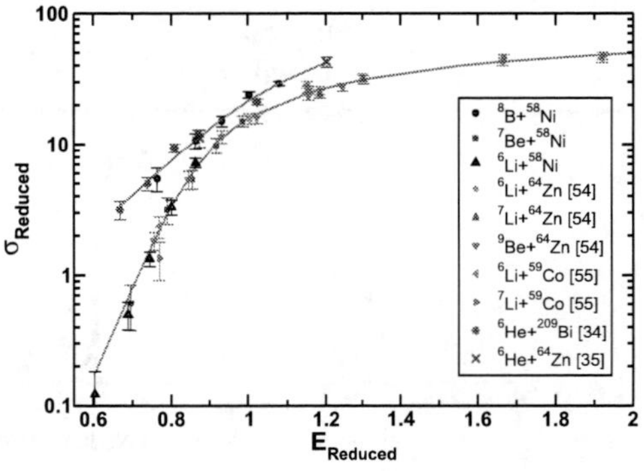

Figure 2. Reduced cross sections from Ref. [4] compared to other data. The curves are to guide the eye.

target masses, all data for halo systems did also fall on a common path, which lies above that for the weakly-bound systems. So, two outstanding features of reactions with halo nuclei can be pointed out from this plot: a) The respective reaction cross sections show a strong enhancement with respect to those corresponding to weakly-bound projectiles, and b) Reactions with the proton-halo projectile ^{8}B show similar enhancement as those with the neutron-halo projectile ^{6}He, despite important differences in structure, binding energy and reaction mechanisms for these two projectiles.

2.2. Characterizing halo, weakly-bound and strongly-bound systems by respective reduced barriers

The previous results suggested a more extensive comparison including further systems [9]. As shown in Fig. 3, additional data for one more halo system, ^{6}He + ^{27}Al, as well as data for a strongly-bound projectile, ^{16}O, were also included. The data for ^{27}Al extended the energy range of the halo systems to higher reduced energies, corroborating the previous trend, while the data for ^{16}O defined a new path, lying further down in the plot. So, in the space of reduced quantities there are three trajectories, one corresponding to halo systems, one for weakly-bound systems, and one for the strongly-bound projectile. These trajectories can be characterized by Wong's model [10] with appropriate barrier parameters. This model is based on the simple idea that a reaction between two interacting nuclei will occur as long as the corresponding Coulomb barrier is passed. In the present situation, a reduced barrier can be defined [11] which serves to characterize each curve. Wong's formula in this case is written as

$$\sigma^W_{Red} = \frac{\epsilon_0 r^2_{0b}}{2E_{Red}} ln\{1 + exp[\frac{2\pi}{\epsilon_0}(E_{Red} - V_{Red})]\}. \quad (1)$$

The reduced radius r_{0b} is simply the barrier radius divided by ($A_p^{1/3}$ + $A_t^{1/3}$), and both the

Table 1. Reduced barrier parameters for each curve in Figs. 3, 4.

Fig.	curve	V_{Red}	r_{0b}	ϵ_0
3	dash (halo projectiles)	0.79	1.79	0.49
	solid (weakly-bound proj.)	0.82	1.64	0.34
	dash-dot ($^{16}O + ^{64}Zn$)	0.87	1.56	0.14
4	solid (^{6}He projectiles)	0.780±0.014	1.79±0.04	0.43±0.06
	dash(^{4}He projectiles)	0.913±0.005	1.39±0.05	0.175±0.006

barrier curvature parameter $\hbar\omega_0$ and the barrier height V_b are reduced in the same way as the energy (see above) to obtain ϵ_0 and V_{Red}, respectively. The curves in Fig. 3 correspond to fits of eq. 1 to the respective experimental points, yielding the parameter values indicated in the first three lines of Table 1.

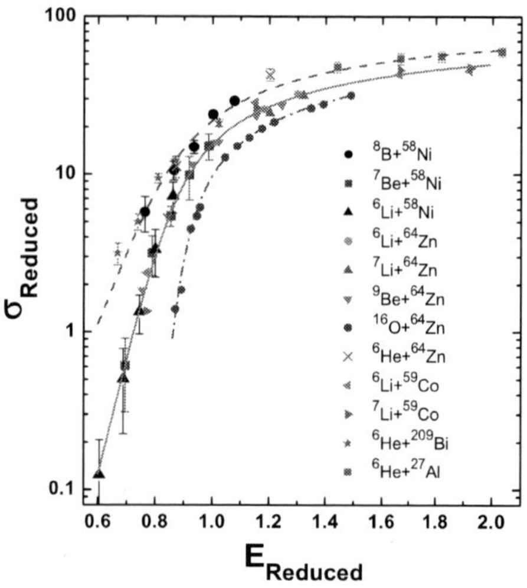

Figure 3. Comparison of reduced cross sections for twelve different systems. Figure adapted from Ref. [9].

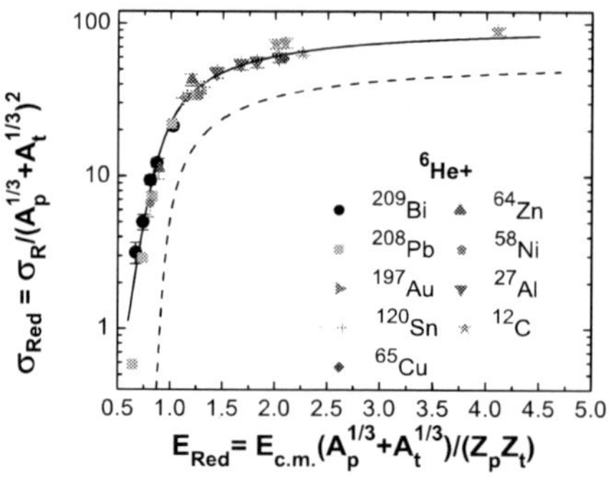

Figure 4. Reduced total reaction cross sections for ^{6}He systems. The solid line is a Wong function fit to the data, while the dashed line is the result of fitting similar data, but for ^{4}He projectiles. Taken from Ref. [3].

Clearly, all the halo systems studied here can be characterized by a single reduced barrier, from which the actual barrier parameters for each particular system can be deduced (an equivalent statement holds for the respective weakly-bound systems). The existence of this systematic behavior suggests that, for a given halo nucleus, the effects of the halo can be investigated by comparing the reduced plots for both, the halo projectile and its core. Since ^{6}He is the halo nucleus for which the largest number of measurements exist, it was only natural to start by doing the mentioned comparison for data corresponding to this neutron-halo projectile, as discussed below.

2.3. Static and dynamic effects of the halo

A comparison between total reaction data for systems with ^{6}He projectiles and similar data for systems where the projectile was the respective core, ^{4}He, was the subject of Ref. [11]. Further discussion on the same topic was included also in Ref. [3]. For the sake of completeness, a brief summary of the main results is given here.

Journal of Physics: Conference Series **387** (2012) 012001 doi:10.1088/1742-6596/387/1/012001

Figure 4 shows total reaction data for a large variety of ^{6}He systems. In addition to the three ^{6}He systems of Fig. 3, data for six other targets were included. As a result, both the mass and the reduced energy ranges for the corresponding systems of the previous figure were considerably extended. A fit of eq. 1 to the more complete data set yielded the barrier parameters shown in line 4 of Table 1, where estimated uncertainties taking into account data spread and respective error bars are also indicated. By comparing lines 1 and 4 of the table, one can conclude that the previously observed trend of the data for halo systems is corroborated in the new analysis, even though the ^{8}B data were excluded here for obvious reasons.

An analysis of experimental data for total reaction cross sections was also done for ^{4}He systems covering the same mass and energy ranges as those for ^{6}He. The respective reduced points (not shown in the figure) do also show a well defined trend, which is quite different from the previous one. The trajectory followed by the ^{4}He systems can also be characterized by a reduced Wong curve, shown by the dashed line in Fig. 4, with parameter values given in the fifth line of Table 1. It can be seen that the reduced cross sections fall systematically lower in this case and the slope for fall-off at low energies is steeper. The differences observed in lines 5 and 3 of Table 1 for the reduced barrier parameters of ^{4}He and ^{16}O, respectively, might be a consequence of the eccentric character that has been observed for the nuclear density of ^{4}He [12], but further study is required to corroborate this point.

A consideration of the big differences in reduced radii for systems with ^{6}He and ^{4}He (lines 4 and 5 of Table 1), lead to the conclusion that the shift of the ^{6}He cross sections to higher values with respect to those corresponding to ^{4}He systems, is consistent with a static effect of the halo. The larger nuclear size associated to the halo would "push" the barrier away to longer distances, at the same time lowering the barrier height [3, 11]. Likewise, the smaller slope shown by the ^{6}He data at low energies is most probably a signature for dynamic effects of the halo. Such effects have been observed experimentally as an enhanced probability of transfer and/or breakup near and below the barrier. Within the context of Wong's model, they would be simulated by an extremely narrow barrier. Indeed, it can be seen from Table 1 that the ϵ_o value for ^{6}He systems is much larger than the one associated to ^{4}He (by a factor of 2.5). This parameter determines the respective barrier curvature and it is also closely related to the low-energy slope of the corresponding curve in Fig. 4. It can be shown that a large value of ϵ_o (or $\hbar\omega_0$) corresponds to an increased diffuseness of the absorption in l space [9], consistent with a higher probability for peripheral processes.

3. σ_{fus}(E) for halo systems

Near-barrier fusion excitation functions, σ_{fus}(E), have been measured for three of the halo systems mentioned above. These are the ^{6}He + ^{209}Bi system [13], the ^{6}He + ^{64}Zn system [14, 15], and the ^{8}B + ^{58}Ni system, which was measured very recently [16]. In this section, a quite interesting common behavior found from these measurements, which is different from the behavior for normal systems, will be pointed out.

Since direct processes are quite important for halo systems at near barrier energies, the fusion cross sections are always lower than the total reaction cross sections, $\sigma_{fus}(E) \leq \sigma_R(E)$ (see Fig. 5). For the latter, we already have a good description in terms of a Wong function with well defined barrier parameters (Table 1). These parameters should be associated not to a bare potential but to some sort of effective interaction potential taking into account all possible interactions. In principle, one could also fit a Wong function to the respective fusion data and the resulting parameters should then be associated instead to an effective fusion potential. These two potentials should be consistent with the same bare potential, connected to it through respective polarization potentials simulating the effects of the involved channel couplings in each case. This subject is presently being investigated and some results will be published elsewhere [17].

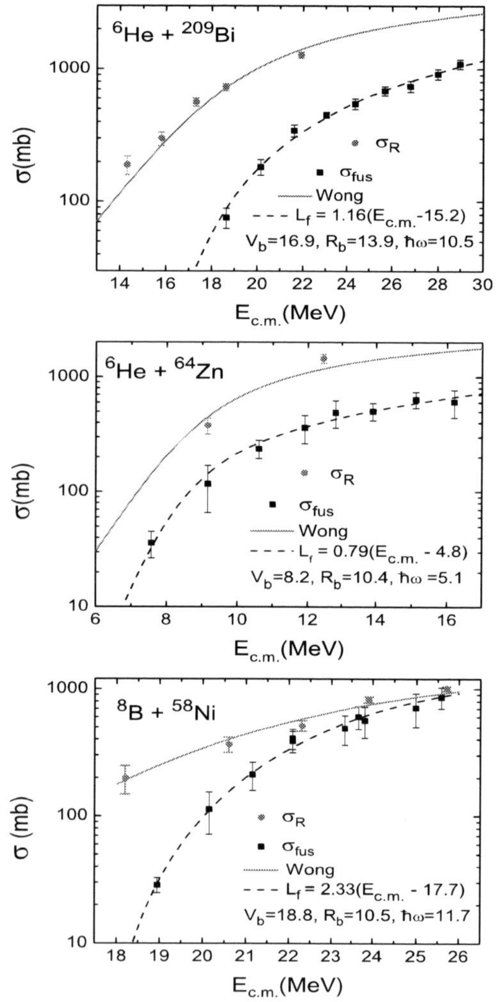

Figure 5. Total reaction σ_R and fusion data σ_{fus} for three halo systems. The solid (dashed) line corresponds to the Wong (linear L_f) model, both of them calculated with the indicated barrier parameters. Adapted from Ref. [18].

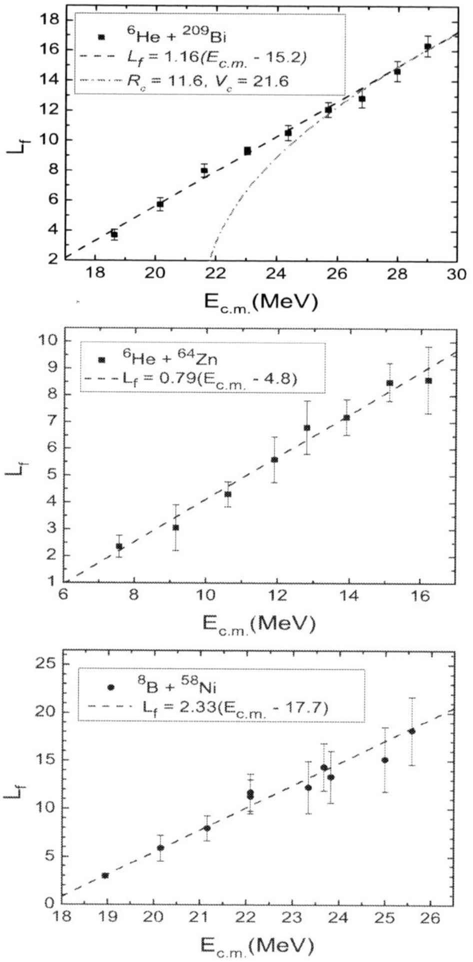

Figure 6. Values of L_f extracted from the fusion data for ^{6}He + ^{209}Bi, ^{6}He + ^{64}Zn and ^{8}B + ^{58}Ni. The dashed lines are linear fits to the respective points while the dash-dotted curve illustrates what would be expected for normal systems. Adapted from Ref. [18].

Alternatively, one may wonder whether for a given system it is possible to describe both the total reaction and the fusion data within the same framework, *i.e.*, by using the same barrier parameters. In fact, it is possible to show that, by using the barrier parameters extracted from the total reaction data, it is possible to describe also the fusion data [18]. Briefly, while the total reaction cross sections are calculated from a sum of terms of the form $(\pi/k^2)(2l+1)T_l$ over all partial waves (T_l are the involved transmission coefficients), the respective fusion cross sections can be estimated from the same sum, but restricted to a maximum angular momentum for fusion L_f. Under this assumption, it was shown in Ref. [18] that the fusion cross section can be approximately calculated by using an analytic function similar to the one obtained in

Journal of Physics: Conference Series **387** (2012) 012001 doi:10.1088/1742-6596/387/1/012001

Wong's model:

$$\sigma_{fus}(E) = [\hbar\omega_0 R_b^2/(2E)]ln[(1+e^{-y_0})/(1+e^{-z})], \tag{2}$$

where

$$y_0 = 2\pi(V_0 - E)/(\hbar\omega_0), \quad and \quad z = y_0 + [\pi\hbar/(\omega_0 \mu R_b{}^2)]L_f(L_f+1). \tag{3}$$

The barrier parameters for each system in Fig. 5 can be obtained from the values given in the first line of Table 1. It is then straightforward to obtain the energy dependence of L_f from the fusion data. For each measured fusion cross section σ_{fus}^{exp}, one can find the L_f value that, when substituted in eq. 2, reproduces σ_{fus}^{exp}. This was done for the three halo systems mentioned above, with the results shown in Fig. 6. All three halo systems show a linear energy dependence for L_f. This is surprising, especially because it turns out to be quite different from what would be expected based on observations for many other systems.

For the case of "normal" systems, if one goes to relatively high energies above the barrier, fusion starts being lower than the total reaction cross section. It was shown a long time ago that a critical distance of approach is the relevant quantity limiting fusion in this case (see discussion in Ref. [18]). In other words, only those nuclei that reach a critical distance R_c get fused. In the L_f plot, the critical distance condition looks like the dash-dotted curve shown in Fig. 6 for the case of the ^{6}He + ^{209}Bi system. Clearly, this can not explain the behavior of the halo system. A critical distance that increases with decreasing energy would be needed in order to reproduce the data points. Therefore, a new mechanism for fusion should be found in order to describe the previous observations for halo systems. It is worth mentioning that weakly-bound systems are probably on the verge of the transition between the behavior predicted by the critical distance model and that of the linear-L_f model [18].

The predictions of the linear-L_f model for the respective fusion excitation functions can be calculated by replacing the corresponding functions L_f(E) (see insets in Fig. 6) in eqs. 2, 3. The obtained description of the experimental data is very good, as shown in Fig. 5 by the dashed lines. It will be interesting to measure near barrier fusion excitation functions for additional halo systems and verify whether a linear-L_f model does also describe the corresponding data.

With respect to the comparison between fusion of neutron-halo vs proton-halo systems, we may mention a qualitative difference in the respective excitation functions for fusion induced by ^{6}He and ^{8}B, which can be observed in Fig. 5. While fusion tends to saturate the total cross section at high energies for the ^{8}B case, the corresponding curves σ_{fus}(E) and σ_R(E) stay rather parallel to each other in the respective plots for the ^{6}He projectile. Apparently, there is a strong fusion enhancement in the proton-halo case which is not present for neutron halos. This remarkable difference, first noticed in Ref. [16] and further discussed in Ref. [18], may result from the different role played by Coulomb polarization in the cases of charged and neutral halos.

This section may be summarized by stating two conclusions: a) By introducing a maximum angular momentum for fusion, L_f, it is possible to describe both the fusion and the total reaction cross sections for halo nuclei within the same framework, *i.e.*, by using the same barrier parameters, and b) A linear energy dependence of L_f seems to be the new ingredient describing the fusion data for halo systems. This simple energy dependence works well for all cases considered, which suggests that this is an important step towards a more detailed understanding of the underlying reaction mechanism.

4. Conclusions

Recent experimental results for total reaction and fusion cross sections of halo nuclei are discussed. A comparison of σ_R with corresponding data for weakly-bound and strongly-bound systems shows three different trajectories in a reduced plot, one for each group. The strong enhancement observed for halo systems can be interpreted in terms of static and dynamic effects of the halo. While static effects are associated to an increased barrier radius and produce an enhancement of σ_R in the whole energy region, dynamic effects are important only in the near barrier region and are simulated by an extremely small barrier width which in turn produces a slower fall off of the data at low energies.

The existing experimental results for fusion of halo systems are discussed by using a simple analytic model which describes the fusion cross sections in terms of a maximum angular momentum for fusion, L_f. It is shown that a linear energy dependence for L_f gives a surprisingly good description of the experimental fusion excitation functions for both proton-halo (^{8}B + ^{58}Ni) and neutron-halo (^{6}He + (^{209}Bi, ^{64}Zn)) systems. This behavior is different from that expected for normal systems so it is possibly a characteristic feature of true halo systems. Finding a fusion mechanism that explains this behavior should help to improve our present knowledge of these exotic nuclei.

Acknowledgments

This work was partially supported by CONACYT, México.

References

[1] Tanihata I et al. 1985 Measurements of interaction cross sections and radii of He isotopes *Phys. Lett.* **B160** 380
[2] Mazzocco M 2010 *Int. J. Mod. Phys.* E **19** 977
[3] Aguilera E F, Kolata J J, Martel I, Sánchez-Benítez A M, and Acosta L 2012 Recent results in reactions using radioactive ion beams *Proc. XXXIV ed. of the Brazilian Workshop on Nucl. Phys. June 2011* (PoS Conference Proceedings) to be published (available on line)
[4] Aguilera E F *et al.* 2009 *Phys. Rev.* C **79** 021601(R)
[5] Lee M Y *et al.* 1999 *Nucl. Instrum. Methods Phys. Res.* A **422** 536
[6] Gomes P R S, Lubian J, Padron I, and Anjos R M 2005 *Phys. Rev. C* **71** 017601
[7] Aguilera E F *et al.* 2000 *Phys. Rev. Lett.* **84** 5058
[8] Aguilera E F *et al.* 2001 *Phys. Rev.* C **63** 061603(R)
[9] Kolata J J and Aguilera E F 2009 *Phys. Rev.* C **79** 027603
[10] Wong C Y 1973 *Phys. Rev. Lett.* **31** 766
[11] Aguilera E F, Martel I, Sánchez-Benítez A M, and Acosta L 2011 Systematics of reactions with ^{4,6}He: Static and dynamic halo effects and evidence for core-halo decoupling *Phys. Rev. C* **83** 021601(R)
[12] Gasques L R, *et al.* 2003 Experimental determination of the surface density for the ^{6}He exotic nucleus, *Phys. Rev. C* **67** 024602
[13] Kolata J J *et al.* 1998 *Phys. Rev. Lett.* **81** 4580
[14] Di Pietro A *et al.* 2004 *Phys. Rev.* C **69** 044613
[15] Di Pietro A *et al.* 2006 *Phys. At. Nucl.* **69** 1366
[16] Aguilera E F *et al.* 2011 *Phys. Rev. Lett.* **107** 092701
[17] Gómez-Camacho A and Aguilera E F 2012 *J. Phys.: Conf. Ser.* **387** 012002
[18] Aguilera E F and Kolata J J 2012 *Phys. Rev.* C **85** 014603

Interaction spatial barrier distributions for the proton-halo system ^{8}B+^{58}Ni

A Gómez Camacho[1] and E F Aguilera

[1]Departamento de Aceleradores, Instituto Nacional de Investigaciones Nucleares, Apartado Postal 18-1027, C.P. 11801, México, D.F.

E-mail: arturo.gomez@inin.gob.mx

Abstract. Radial barrier distributions for the proton-halo system ^{8}B+^{58}Ni are obtained by the use of Woods-Saxon polarization potentials calculated from a fit to elastic scattering angular distribution data. The strength, diffuseness and reduced radii of the polarization potentials, besides fitting the data, are such that reproduce the barrier height and position as extracted from a Wong´s fit to the universal behavior of reduced interaction cross sections for halo systems. It is found that, for the proton-halo system ^{8}B+^{58}Ni, the interaction barrier height V_I and the interaction barrier radius R_I result correspondingly lower and larger than the Coulomb nominal barrier height V_B and barrier position R_B as expected from Barrier Penetration Models and double folding density dependent interactions such as the São Paulo Potential. It is also shown that, the determined energy dependent polarization potentials, besides satisfying the dispersion relation, present a behavior consistent with the Breakup Threshold Anomaly.
Keywords. Nuclear Reactions, weakly bound nuclei, threshold anomaly. PACS Numbers: 25.70.Bc,24.10.Ht,27.20.+n

1. Introduction

Recently, elastic scattering measurements[1] of ^{8}B and ^{7}Be on a ^{58}Ni target were shown to provide strong evidence for proton-halo effects on ^{8}B reactions. It was established that a characteristic feature expected for any halo state, such as ^{8}B or ^{6}He, an underlying decoupling between the core and the valence nucleons occurs. That is, the ^{8}B (^{6}He) reaction cross section could be entirely accounted for, by breakup (transfer) of the halo state (p or $2n$) plus reactions that occur with the ^{7}Be (^{4}He) core[2, 3, 4]. It was also pointed out that, in the space of reduced cross sections and reduced energies, the interaction cross sections for ^{8}B (proton halo) and ^{6}He (neutron halo) with different targets, lie on the same trajectory. Similarly, "normal" weakly bound (non halo) and tightly bound systems follow characteristic paths. That is, in the space of reduced cross sections defined by,

$$\sigma_{red} = \frac{\sigma_R}{\eta}, \; \eta = (A_p^{1/3} + A_T^{1/3})^2, \tag{1}$$

where A_P and A_T are the mass numbers of the projectile and target nuclei, and σ_R is the reaction cross section, and reduced energies given by,

$$E_{red} = \frac{E_{c.m.}}{\xi}, \; \xi = \frac{Z_p Z_T}{\eta^{1/2}}, \tag{2}$$

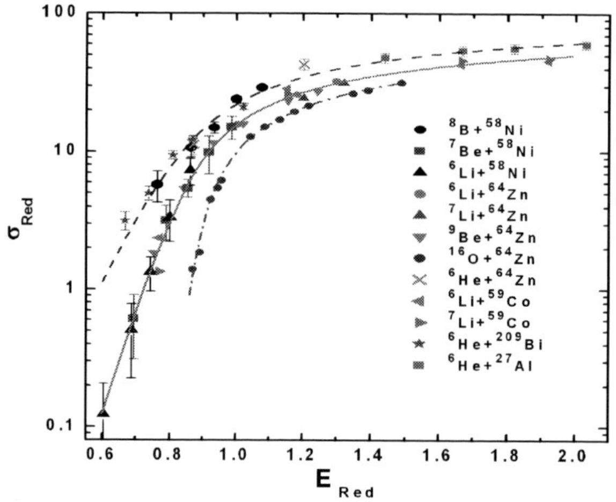

Figure 1. Reduced total reaction cross sections for a number of systems. Data for $^8B+^{58}Ni$, $^7Be+^{58}Ni$, $^6Li+^{58}Ni$ from Ref.[1], $^{6,7}Li+^{64}Zn$, $^9Be+^{64}Zn$, $^{16}O+^{64}Zn$ from Ref.[5], $^6He+^{64}Zn$ from Ref.[6], $^{6,7}Li+^{59}Co$ from Ref.[7], $^6He+^{209}Bi$ from Ref.[8], and $^6He+^{27}Al$ from Ref.[9].

where Z_P and Z_T are the charges of the nuclei. Reactions with several types of projectiles describe different behaviors regardless of the nuclear structure, binding energies or reaction mechanisms. This feature is shown in Fig.1, for a number of systems. It is seen that for an energy range from 0.68 MeV$< E_{red} <$ 2.1 MeV, the trajectory described by the halo systems $^8B+^{58}Ni$ and $^6He+^{209}Bi$ lies above that for "normal" weakly bound and, this in turn, is above that for the tightly bound system $^{16}O+^{64}Zn$. Thus, it could be established that, if the trajectories of Fig.1 should be characterized by the mean value of the projectile binding energies, then these trajectories lie progressively lower as the binding energy increases. Also shown in Fig.1 are Wong model fits illustrated by the curves to the three different trajectories as described by Kolata *et al.*,[2]. The Wong model[10] gives a simple expression for the interaction cross section as a function of the energy $E_{c.m.}$ in terms of the s-wave interaction barrier. it depends on three parameters, namely the barrier height V_0, the barrier position R_0 and the curvature $\hbar\omega_0$, *i.e.*,

$$\sigma = (R_0^2\hbar\omega_0/2E_{c.m.})\ln\{1+\exp\left[2\pi(E_{c.m.}-V_0)/\hbar\omega_0\right]\}, \tag{3}$$

where the interaction barrier radius R_0 is defined by $R_0 = r_{0b}(A_P^{1/3}+A_T^{1/3})$. Thus, the Wong model fit for a given system consists of finding the parameters that best adjust the experimental reaction cross section data. In terms of reduced parameters, Eq.(3) can be written as[4],

$$\sigma_{red} = (r_{0b}^2\epsilon_0/2E_{red})\ln\{1+\exp\left[\frac{2\pi}{\epsilon_0}(E_{red}-V_{red})\right]\}, \tag{4}$$

where the reduced curvature, barrier position and barrier height are respectively given by,

$$\epsilon_0 = \hbar\omega_0/\xi,\ r_{0b} = R_0\eta^{-1/2},\ V_{red} = V_0/\xi, \tag{5}$$

where ξ and η are given by Eqs.(1) and (2). As discussed in Ref.[4], the Wong-model fit for halo-systems shown by the dashed-line of Fig.1 is achieved with the reduced parameters, $V_{red} = 0.78 \pm 0.014$ MeV, $r_{ob} = 1.79 \pm 0.04$ fm and $\epsilon_0 = 0.43 \pm 0.06$ MeV. It is very interesting to point out that, these reduced parameters were determined by fitting experimental reduced reaction cross section data of the neutron-halo projectile ^6He with a variety of targets, measured by different authors and by different technics. It is remarkable that the experimental reduced reaction cross section data for the proton-halo ^8B with ^{58}Ni extracted from the measurements

Journal of Physics: Conference Series **387** (2012) 012002 doi:10.1088/1742-6596/387/1/012002

of Ref.[1], is also well described by the same dashed-curve of Fig.1. Thus, from the "universal" halo-reduced parameters just obtained, *i.e.*, V_{red}, r_{ob} and ϵ_0, the corresponding Wong-model parameters of Eq.(3), for the interaction barrier of the ^{8}B+^{58}Ni system are,

$$V_I = 18.6 \pm 0.33\ MeV, R_I = 10.5 \pm 0.23\ fm,\ \hbar\omega_I = 10.25 \pm 0.2\ MeV. \tag{6}$$

It is interesting to notice that, for this nuclear system, the interaction barrier height V_I and barrier position R_I are correspondingly lower and larger, than the Coulomb barrier height and barrier position, as determined by Barrier Penetration Models (BPM)[11] or double-folding density dependent potentials like the São Paulo Potential (SPP)[12, 13], that predict $V_B = 20.8$ MeV and $R_B = 8.89$ fm. However, it is important to notice, that when the polarization potential is included, the interaction barrier can be modified by the action of this potential. A possible explanation is twofold, namely, the low threshold energy of ^{8}B (0.137 MeV) that facilitates the production of breakup reactions ($^8B \rightarrow p + ^7Be$) and thus lowers the barrier. And the large radial density distribution of the halo structure of ^{8}B, whose effect could be of "pushing" the nuclear potential to larger distances and thus the barrier. These two effects can be due to the action of the polarization potential, that accounts for couplings between the elastic channel and the different absorption mechanisms, *i.e.*, fusion, breakup and inelastic reactions. It is the purpose of the present work to calculate the interaction barrier V_I and barrier position R_I, by calculating the effect that the polarization potential $\mathcal{V}_{pol}$ has on the barrier. This potential has real and imaginary parts, $\mathcal{V}_{pol} = V_{pol} + iW_{pol}$, where the real part V_{pol} results from virtual excitations to the non-elastic channels and W_{pol} describes the loss of incident flux. The energy dependent real and imaginary potentials $V_{pol}(E)$ and $W_{pol}(E)$ are linked by the dispersion relation[14, 15], and should be such that comply with the Breakup Threshold Anomaly (BTA) as proved be the case for many weakly bound systems.

The real and imaginary parts of the polarization potential are chosen to have Woods-Saxon volumetric geometrical shapes with the same diffuseness and reduced radii parameters. The potential parameters, *i.e.*, the diffuseness a_{pol}, reduced radius r_{pol} and the strengths $V_{0,pol}$ and $W_{0,pol}$ are determined from fittings to elastic scattering angular distributions of Ref.[1] at the energies $E_{c.m.} = 18.17, 20.56, 22.23, 23.9$ and 25.75 MeV from which reaction cross sections are extracted. The real part of the polarization potential V_{pol}, should be such that, when added to the bare potential V_{bare} of the type SPP of Ref.[12, 13], correctly gives the interaction barrier height and barrier position. It is expected that the calculated values of the parameters a_{pol} and r_{pol} can be related to the large radial extension of the halo projectile ^{8}B and be such that push the barrier to higher distances as predicted by the Wong model.

In the next section, the basic features of the model used are presented, section 3 is dedicated to the results of the calculations and finally a summary is given.

2. Brief Model Description.

The Hamiltonian H for the nuclear system is of the form,

$$H_a = T_a + V_a, \tag{7}$$

where the distorted wave $\chi_a^{(+)}$ in the incident elastic channel a satisfies the expression,

$$(T_a + V_a)\chi_a^{(+)} = E_a\chi_a^{(+)}. \tag{8}$$

The potential V_a is defined by,

$$V(r,E) = V_{Coul}(r) - V_{bare}(r,E) - \mathcal{V}_{pol}(r,E), \tag{9}$$

where the subindex a has been suppressed. Here, $V_{bare}(r,E)$ is a double-folding density dependent nuclear potential (São Paulo Potential).

$$V_{bare}(r,E) = V_{SPP}(r,E) = V_F(r)e^{-4v^2/c^2}. \tag{10}$$

where,

$$v^2(r,E) = \frac{2}{\mu}\left[E - V_{Coul}(r) + V_{bare}(r,E) + \mathcal{V}_{pol}(r,E)\right], \tag{11}$$

is the relative velocity of the colliding nuclei and,

$$V_F(r) = \int \rho(\mathbf{r_1})\rho(\mathbf{r_2})V_0\delta(\mathbf{r}-\mathbf{r_1}+\mathbf{r_2})d\mathbf{r_1}\mathbf{dr_2}. \tag{12}$$

An extensive systematics of nuclear densities of the SPP has been proposed by L.C. Chamon *et al.*,[12, 13], that is, within the two-parameter Fermi distribution approach to describe the densities, the radii of the distributions are given by,

$$R_{density} = 1.31A^{1/3} - 0.84, \tag{13}$$

where A is the number of nucleons in the nucleus. In a zero-range approximation the strength $V_0 = -456$ MeV-fm^3 is assumed. Also, within the systematics, the matter diffuseness a of the density distributions has an average value $a = 0.56$ fm throughout the periodic table. The normalization of the double-folding potential e^{-4v^2/c^2}arising from the Pauli non-locality due to exchange of the nucleons vanishes at near barrier energies, and therefore the São Paulo potential becomes an usual double-folding potential.

$\mathcal{V}_{pol}(r,E)$ of Eq.(9), is defined by $\mathcal{V}_{pol}(r,E) = V_{pol}(r,E) + iW_{pol}(r,E)$, where the real and imaginary potentials V_{pol} and W_{pol}, are Woods-Saxon volume potentials with the same reduced radii r_{VW} and diffuseness a_{VW} parameters, *i.e.*,

$$V(W)_{pol}(r,E) = \frac{V(W)_{pol,0}(E)}{1+\exp[(r-R_{VW})/a_{VW}]}, \tag{14}$$

where $R_{VW} = r_{VW}(A_p^{1/3} + A_T^{1/3})$. The potential strengths $V_{pol,0}(E)$, $W_{pol,0}(E)$, the reduced radius r_{VW} and the diffuseness a_{VW} are determined during the fitting of elastic scattering angular distributions for ^{8}B+^{58}Ni. It is expected that the polarization potentials also satisfy the dispersion relation[14, 15],

$$V_{pol}(R_{sa},E) = \frac{1}{\pi}\mathcal{P}\int_0^\infty \frac{W_{pol}(R_{sa},E')}{(E'-E)}dE', \tag{15}$$

at the strong absorption radius R_{sa}. Finally, once the parameters of the polarization potentials are obtained, the interaction barrier V_I will be given by,

$$V_I(r,E) = V_{Coul}(r) - V_{bare}(r,E) - V_{pol}(r,E), \tag{16}$$

3. Results.

The bare potential of Eq.(10) is a parameter-free potential, that is, for a given nuclear system, it depends only on the incident energy. Nuclear matter densities are calculated with matter radii and diffuseness determined by the systematics. The Coulomb barrier calculated with the bare potential V_{bare} gives a barrier height and barrier position of approximately 20.8 MeV and 8.89 fm respectively for all the energies for which elastic scattering angular distributions were measured[1]. In order to "push" the interaction barrier to larger distances as required by Eq.(6),

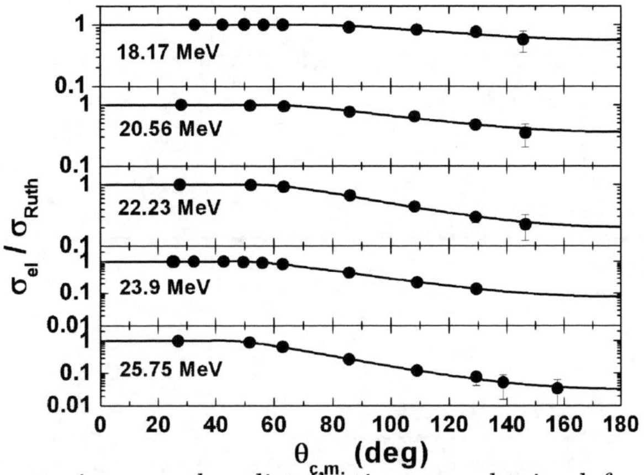

Figure 2. Elastic scattering angular distributions as obtained from the analysis of elastic scattering angular distributions of ^{8}B+^{58}Ni. Data from Ref.[1].

a large reduced radius parameter r_{VW} of the polarization potentials V_{pol} and W_{pol} is needed. The values of the parameters of Eqs.(14), resulting from a χ^2-analysis of elastic angular distributions of ^{8}B+^{58}Ni are listed in Table 1 and the results are shown in Fig.2. Table 1, also shows the calculated $\sigma_{R,the}$ and experimental $\sigma_{R,\exp}$ reaction cross section as extracted from the angular distributions.

Table 1. Woods-Saxon potential parameters obtained from a fitting of elastic scattering angular distributions for ^{8}B+^{58}Ni of Ref.[1]. Potential strengths in MeV, diffuseness reduced radii in fm. Reaction cross sections in mb.

$E_{c.m.}$	$W_{pol,0}$	$V_{pol,0}$	a_{VW}	r_{VW}	χ^2/N	$\sigma_{R,\exp}$	$\sigma_{R,the}$
18.17	89.1	12.1	0.32	1.67	0.10	198 ± 10	198.6
20.56	31.7	10.2	0.26	1.67	0.54	365 ± 20	367.8
22.23	18.3	8.3	0.25	1.66	0.40	515 ± 20	518.9
23.90	18.1	4.1	0.30	1.66	0.73	827 ± 10	825.2
25.75	18.0	5.6	0.33	1.66	0.71	1007 ± 40	1004.6

From eqs.(14) and (16), the interaction barrier can now be calculated for the different energies. In Fig.3, $V_I(r,E) = V_{Coul}(r) - V_{bare}(r,E) - V_{pol}(r,E)$ is shown for $E_{c.m.} = 18.17$ MeV. As seen, due to V_{pol}, the interaction barrier height and position are much closer to the values of Eq.(6), obtained from reduced cross sections. Fig.4, shows the results for the other energies, for which similar values are obtained.

It is important to point out, that V_{pol} includes the effects on the barrier, from all the possible reaction channels, that is, the attractive polarization effect due to couplings to the fusion channel, the attractive effect from inelastic couplings and the repulsive contribution from breakup processes. However, all of these absorption couplings, contribute to the total reaction as attractive potentials (minus sign in Eq.(16)). Of course, the fusion barrier, should include the separate attractive or repulsive effects from the different polarization potentials with their respective signs.

The Threshold Anomaly present in reactions with tightly bound nuclei, is deduced by the correlated behavior between the energy dependent potentials $V_{pol}(r,E)$ and $W_{pol}(r,E)$. That is, as the collision energy between the nuclei is decreased towards the Coulomb barrier V_B, the absorptive imaginary potential $W_{pol}(r,E)$ also decreases, a fact that is related to the closure

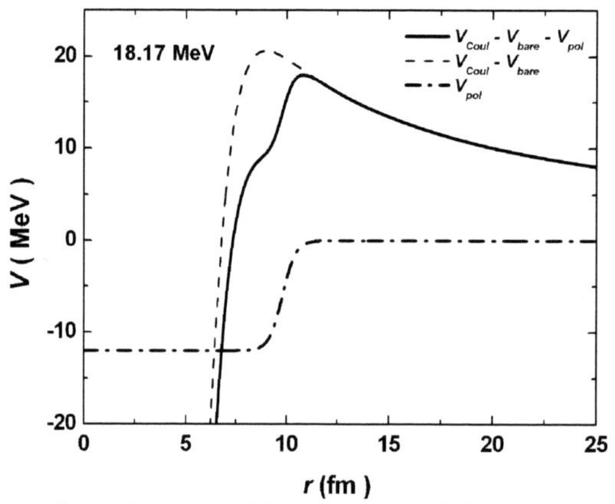

Figure 3. Barrier height and position at $E_{c.m.} = 18.17$ MeV as obtained with the total nuclear potential, $V_I(r, E) = V_{Coul}(r) - V_{bare}(r, E) - V_{pol}(r, E)$.

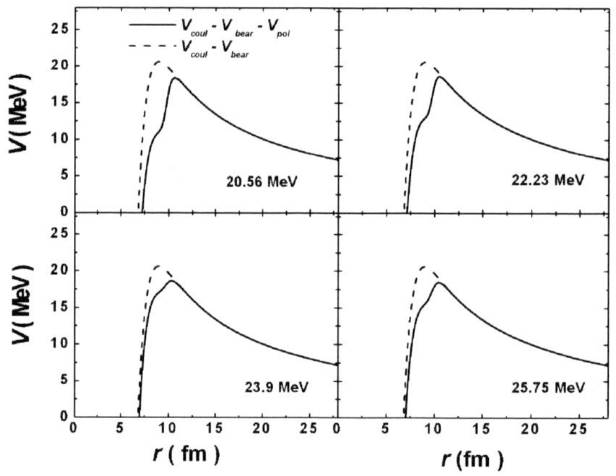

Figure 4. Barrier height and position at $E_{c.m.} = 20.56, 22.23, 23.9$ and 25.75 MeV with corresponding $V_I(r, E) = V_{Coul}(r) - V_{bare}(r, E) - V_{pol}(r, E)$.

of reaction channels. On the other side, the real potential, $V_{pol}(r, E)$ shows an increasing-decreasing behavior (bell-shape) just around the barrier, producing a more attractive potential, when added to the bare nuclear potential. Thus, the net effect of V_{pol} is to lower the Coulomb barrier and consequently fusion is enhanced. Recent studies show that this is not the case, when weakly bound nuclei are involved[16, 17, 18]. Instead, weakly bound projectiles, with low threshold energies produce strong couplings between the breakup/transfer and elastic channels. That is, for energies around and below the Coulomb barrier, reactions with weakly bound nuclei show strong direct reaction yields. For this reason the imaginary potential $W_{pol}(r, E)$ can not sharply decrease, as it does for tightly bound nuclei. The real polarization potential $V_{pol}(E)$ should include contributions from polarization effects coming from all the possible reaction mechanisms. That is, the attractive polarization effects from fusion and inelastic couplings that lower the fusion barrier and, the repulsive contribution from couplings to the breakup channel. Hence, the former effects enhance fusion while the latter suppresses it. For this reason the energy behavior of the net $V_{pol}(E)$ has not the bell-shape form around the Coulomb barrier as

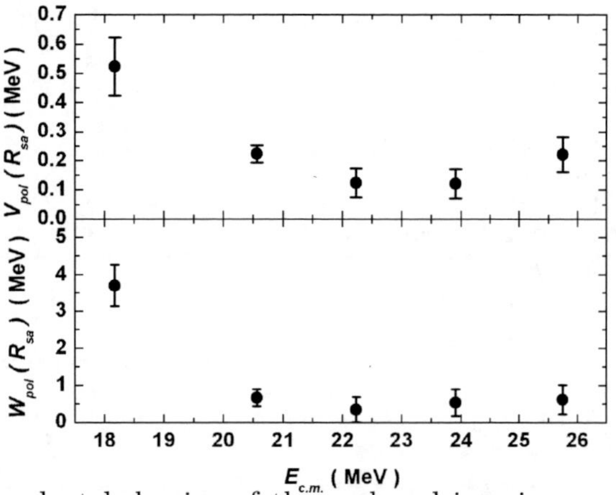

Figure 5. Energy dependent behavior of the real and imaginary polarization potentials V_{pol} and W_{pol} at the strong absorption radius for ^{8}B+^{58}Ni.

for the case of tightly bound nuclei. Instead, as the energy goes below the Coulomb barrier, $V_{pol}(E)$ becomes dominated by the increasing repulsive contribution from breakup effects. This correlated behavior between the real and imaginary polarization potentials is known as the Breakup Threshold Anomaly (BTA)[19]. It is expected that the BTA also be present for halo systems as is the case for ^{8}B+^{58}Ni as seen in Fig.5. Here, $V_{pol}(E)$ and $W_{pol}(E)$ are the strengths of the polarization potentials evaluated at the strong absorption radius R_{sa}. It is clear that $W_{pol}(E)$ does not decrease even for energies lower than the Coulomb barrier, indicating the reaction channels remain open. On the other hand, the increasing strength of $V_{pol}(E)$ manifests that the rising of the barrier due to the breakup couplings are stronger as the energy is lowered.

In summary, the total reaction barrier height and position of the system ^{8}B+^{58}Ni is calculated by means of Woods-Saxon polarization potentials whose parameters are calculated by a χ^2-analysis of elastic scattering data. It has been found that, when these polarization potentials are added to a bare double-folding density dependent nuclear potential (São Paulo Potential), the barrier height and position of the total nuclear potential agree with the values extracted from the Wong´s fit to reduced total cross sections and energies for halo systems. The effect of the polarization potentials is to displace the interaction barrier to a higher position from the usual Coulomb barrier position. This fact should be related to the spatial extension of the proton-halo structure in ^{8}B that pushes the interaction barrier to larger distances. On the other hand, the interaction barrier height (18.6 MeV) is smaller than the corresponding Coulomb barrier height (20.8 MeV). This result should be connected to the small breakup threshold energy of ^{8}B (0.137 MeV), that causes breakup reactions $^8B \rightarrow ^7 Be + p$ to begin at large distances and to continue being appreciable even at small collision energies. This last fact, is also observed in the presence of the Breakup Threshold Anomaly in the reaction ^{8}B+^{58}Ni.

4. Acknowledgments

A. Gómez Camacho and E.F. Aguilera acknowledge the financial support of CONACYT, México.

References

[1] Aguilera E.F. *et al.*, 2009 *Phys. Rev.* C **79** 021601
[2] Kolata J.J. and Aguilera E.F., 2009 *Phys. Rev.* C **79** 027603
[3] Aguilera E.F., Kolata J.J. and Acosta L., 2010 *Phys. Rev.* C **81** 011604 (R)
[4] Aguilera E.F., *et al.*, 2011 *Phys. Rev. C* **83** 021601 (R)

[5] Gomes, P.R.S., *et al.*, 2005 *Phys. Rev.* C **71** 034608
[6] Di Pietro A., *et al.*, 2004 *Phys. Rev.* C **69** 044613
[7] Beck C., Keeley N., Diaz-Torres A., 2007 *Phys. Rev.* C **75** 054605
[8] Aguilera E.F., *et al.*, 2001 *Phys. Rev.* C **63** 061603 (R)
[9] Benjamin E.F., *et al.*, 2007 *Phys. Rev.* C **63** 061603 (R)
[10] Wong C.Y., 1973 *Phys. Lett.* B **647** 30
[11] Vaz L.C. Vaz, Alexander J.M. and Satchler G.R., 1981 *Phys. Rep.* **69** 373
[12] Chamon L.C., *et al.*, 1997 *Phys. Rev. Lett.* **79** 5218
[13] Chamon L.C., *et. al.*, 2002 *Phys. Rev.* C **66** 014610
[14] Mahaux C., Ngô H. and Satchler G.R., 1986 *Nucl. Phys.* A **449** 354
[15] Satchler, G.R., 1987 *Nucl. Phys.* A **472** 591
[16] Gómez Camacho A., Gomes P. R. S. and Lubian J., Journal of Physics, Conference Series, **322** (2011) 012008.
[17] Gómez Camacho A., *et al.*, 2011 *Phys. Rev.* C **84** 034615
[18] Gómez Camacho A., *et al.*, 2010 *Phys. Rev.* C **82** 067601
[19] Hussein M.S., *et al.*, 2006 *Phys. Rev.* C **73** 044610

Journal of Physics: Conference Series **387** (2012) 012003 doi:10.1088/1742-6596/387/1/012003

High precision measurements for the rp-process

G P A Berg [1], Y Fujita [2], J Görres [1], MN Harakeh [3], K Hatanaka [4], A Long[1], R Neveling [5], F D Smit [5], R Talwar [1], A Tamii [4], and M Wiescher[1]

[1] Department of Physics and the Joint Institute for Nuclear Astrophysics, University of Notre Dame, Notre Dame, IN 46556, USA
[2] Department of Physics, Osaka University, Toyonaka, Osaka 560-0043, Japan
[3] Kernfysisch Versneller Instituut, Rijksuniversiteit Groningen, Zernikelaan 25, 9747 AA Groningen, The Netherlands
[4] Research Center for Nuclear Physics, Osaka University, Ibaraki, Osaka 560-0047, Japan
[5] iThemba Laboratory for Accelerator Based Science, Somerset West 7129, South Africa

E-mail: gpberg@bergs.com

Abstract. The explosive nuclear burning of hydrogen at high temperatures and densities on the surface of accreting neutron stars, known as the rp-process, gives rise to a number of observable phenomena related to X-ray bursts. Recent astronomical observations provide unprecedented information, e.g. on atomic abundances in ejecta and time structure in X-ray bursts. To interpret these data requires an understanding of the nuclear processes during the explosive events and, therefore, information on the structure of unstable, proton-rich nuclei. Network model calculations show that the dominant burning processes, after breakout from the hot CNO cycles at sufficiently high temperatures, proceed via proton- and α-induced reactions. Since the reaction rates are very sensitive to the nuclear structure, shell- and statistical-model calculations are often insufficient in predicting exact reaction paths. Therefore, we have been conducting experiments using high-resolution spectrometers to search for resonances that play a role in the rp-process and determine the nuclear-structure, most importantly excitation energies, accurately above the particle thresholds. The techniques and examples of experimental results will be presented and discussed. The status and outlook of direct measurements of stellar reactions rates in inverse kinematics using planned recoil separators will be outlined.

1. Introduction

Cataclysmic nuclear explosions on the surface of accreting white dwarfs are so violent and frequent that they were already observed more than 2000 years ago by ancient Chinese observers with a keen eye on the sky. Since the light from these explosions seemed to appear and disappear, they called them guest stars. In the Middle Ages even the Europeans, with the strong belief that the heavens were eternal and unchangeable, had to admit that new stars seemed to appear suddenly and, therefore, called them novae. Once this fact was accepted, the search and regular observations of novae, as they are still called today, was continued by professional and amateur observers. An impressive example is the recording of the visual light curve of the recurring dwarf nova SS Cygni starting 1896 and continuing to this day by amateurs and published by the AAVSO [1] despite the enormous progress we have made in the instrumental observation in the last decades. Today, we know that the nucleosynthesis of the isotopes, starting soon after the Big Bang, is still happening at many locations of our universe. A number of distinct, mostly exothermic, nuclear reactions is driving these processes. Novae are, now, interpreted as thermonuclear explosions on the surface of accreting white dwarfs in binary stars, but with the advent of satellite based X-ray spectrometers in the 1970s, a second class of much more violent explosions was discovered (now we have about 100 of these sources with multiple eruptions), which emit a strong X-ray signal and are, therefore, named X-ray bursts. These events are associated with accretion processes onto the surface of a neutron star. While novae explosions are driven primarily by the hot CNO cycle, X-ray bursts are driven by the rp-process. The rp-process was first described in 1981 by Wallace and Woosley [2] and identified as a rapid series of (p,γ) reactions and β decays. At higher temperatures, the competing (α,p) reaction (αp-process) can by-pass specific reaction paths. We will give examples how high-resolution studies of reactions can determine details of their reaction paths and help to better understand a variety of recent astronomical observations. When employed at the new generation of high-intensity radioactive beam facilities, recoil separators can directly measure the reaction rates on radioactive target nuclei that are involved in the rp-process.

2. The rp-process and rate calculations

The rp-process occurs at high temperatures (> 0.4 GK) and large densities (>10^4 g/cm^3), following the break out from the hot CNO cycles and continues rapidly, hence the name rapid-proton capture process or rp-process, along energetically possible reaction paths producing a series of proton-rich nuclei up to iron and beyond. Since the β decay of the T = 1 nuclei (^{22}Mg, ^{26}Si, ^{30}S, and ^{34}Ar) in the path have life times of the order of 1 s to 4 s, the explosive event is briefly delayed at these "waiting points" unless the reaction rate of a fast (p,γ) reaction by-passes a particular waiting point.

The conditions, where the rp-process can occur, are found at several locations in the cosmos, like in the hydrogen-accreting surface of neutron stars (X-ray bursts) of a close binary system with high-temperatures and densities. In case of a close binary-star system, the lighter companion is a red giant filled up to the first Lagrange point, where mainly hydrogen spills over into the gravity-potential well of the heavier, but smaller main star at typical rates of 10^{-8} solar masses per year, forming an accretion disk where the material slowly descends onto the surface losing angular momentum in the process. In novae, the most revealing situation arises when the plane of the binary system and the earth are nearly in the same plane leading to eclipses with characteristic timing marks, due to the entry and exit points of both stars and the bright spot, where the hydrogen plunges into the accretion disk.

In an X-ray burster, the hydrogen accumulates on the surface of the neutron star until the temperature and density are high enough for a thermonuclear ignition. Since the conditions are highly electron degenerate the released nuclear energy goes directly into a rapid increase in temperature, igniting more reactions and leading to a thermonuclear runaway.

The rate of reactions with narrow, non-overlapping resonances can be calculated as the sum over individual resonances according to the narrow-resonance formalism outlined in Ref. [3].

$$N_A\langle\sigma v\rangle_{res} = 1.54\times10^5(\mu T_9)^{-3/2}\sum_i(\omega\gamma)_i \times exp(-11.605E_i/T_9)\ [\mathrm{cm^3s^{-1}mol^{-1}}], \quad (1)$$

with μ the reduced mass of the target and projectile in units of amu, T_9 the temperature in units of GK, $(\omega\gamma)_i$ the strength of the ith resonance in units of eV, and E_i the energy of the ith resonance in units of MeV.

If level densities are high the reaction rates can be calculated within the statistical model (Hauser-Feshbach formalism) by using computer codes like CIGAR [4], NON-SMOKER [5] or

Journal of Physics: Conference Series **387** (2012) 012003 doi:10.1088/1742-6596/387/1/012003

TALYS [6]. To calculate the sequence of reaction paths all isotopes that are produced and destroyed have to be included in network calculations [7]

Equation 1 shows that the spectroscopic information of all nuclear levels involved enters in these calculation, but the rates depend exponentially and, therefore, most strongly on the resonance energy. Uncertainties in the resonance energies of the order of 100 keV can result in uncertainties of several orders of magnitude in the calculated rates. Present network calculations take as input measured parameters of single resonances where available. Otherwise, shell-model predictions with typical energy uncertainties of 100 keV or more are used. Only where level densities are high, calculations are performed using the Hauser-Feshbach statistical model. This assumption may not be always correct, in particular if selectivity reduces the level density as, e.g., in the (α,γ) or (α,p) reactions.

Whenever possible, excitation functions of the (p,γ) and (α,γ) reactions are measured directly down to the lowest possible beam energy. However, low cross sections and beam-related background do not allow to measure at energies low enough to correspond to stellar temperatures and extrapolations using the R-matrix approach are necessary [8]. Moreover, this method requires stable target nuclei, and is, therefore, not available for studies of the rp-process involving unstable proton-rich nuclei. Therefore, the above-mentioned method of calculating the reaction rates on the basis of spectroscopic information of the resonance states is applied.

The contributing resonances in the compound nucleus of the stellar reaction are located within the Gamow window [3] typically > 1 MeV above the p- and α-thresholds depending on the stellar temperature. They show up as peaks in the spectra of the final nucleus in a suitable reaction. These reactions can be any neutron-pickup reaction with targets of the most proton-rich stable nuclei. Suitable reactions are $2n$ pickup reactions, e.g., $(^4\mathrm{He},^6\mathrm{He})$ (Ref. [9]),(p,t) (Ref. [10]), $3n$ pickup reaction, e.g., $(^3\mathrm{He},^6\mathrm{He})$ or in special cases the $4n$ pickup reaction $(^4\mathrm{He},^8\mathrm{He})$ [11].

3. Experimental Method

High-resolution spectrometers may be used to resolve and identify resonances that may be located at excitation energies up to about 12 MeV. Well-suited spectrometers are the Grand Raiden spectrometer [12, 13], at the Research Center for Nuclear Physics (RCNP) of the Osaka University in Osaka, Japan and the K600 spectrometer at iThemba LABS, Somerset West, South Africa [14]. Both spectrometers have dispersion-matched beam lines and are equipped for measurements at and near 0°.

Figure 1 shows the cyclotron facility of RCNP including the injector AVF cyclotron, the separated-sector Ring cyclotron, the dispersion-matched WS beam line, and the Grand Raiden spectrometer, consisting of the high-resolution spectrometer used in these studies and the Large Angle Spectrometer (LAS) not used in the present studies.

Grand Raiden has a high-resolving power of $\mathrm{p}/\Delta\mathrm{p} = 37000$ for a 1 mm target spot and is equipped with a fully dispersion-matched beam line. This includes the lateral-dispersion matching to be able to achieve the highest-possible spectral resolution despite the fact that the beam spread is about a factor of 10 worse compared to the resolving power. The beam line also allows for angular-dispersion matching to allow for the reconstruction of the scattering angle with a precision of 0.4° - 0.6° at the target from measurements in the focal plane. This allows the determination of the angular distribution and hence the angular momentum for measurements at 0° within the angle acceptance of the spectrometer of $\pm$ 1° horizontally and $\pm$ 3° vertically. Also, the scattering angle of each event is necessary to achieve the highest-possible resolution by correcting the kinematic effect that changes the energy of the detected reaction product with angle. Typical examples are, e.g., for 200 MeV protons and α-particles a beam spread of about 80 - 150 keV, while the spectrometers allow spectral resolutions of 10 - 20 keV. The difference in the kinematical shift of nuclei involved in the rp-process typically exceed these resolutions and need to be corrected using the measured scattering angle.

Journal of Physics: Conference Series **387** (2012) 012003 doi:10.1088/1742-6596/387/1/012003

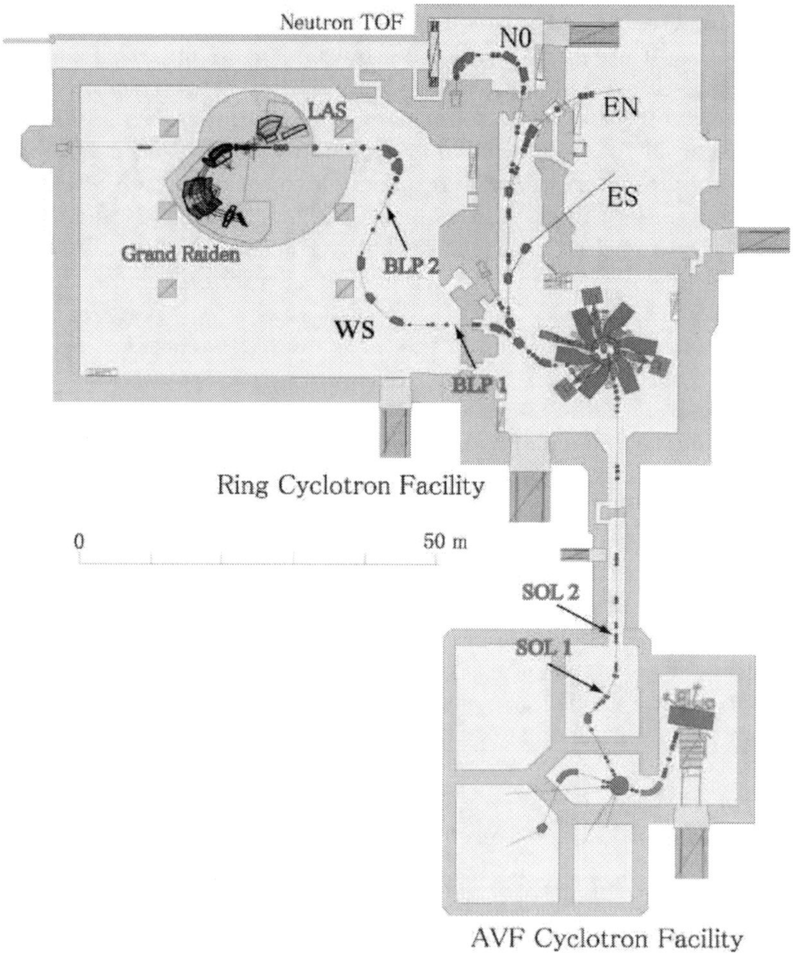

Figure 1. Layout of the RCNP cyclotron facility showing the Grand Raiden spectrometer and the fully dispersion-matched WS beam line.

The highest possible resolution is not only important to separate close resonances but also to determine the excitation energies better than 10 keV. As mentioned above the main reason for the uncertainties of the calculated reaction rates are the uncertainties of the excitation energies. Our studies have shown [10] that this error can be made small compared to uncertainties from the resonance strength and spin assignments if the excitation energy can be determined with an error smaller than 2 - 5 keV. The statistical error of a peak position is given by the width of the peak divided by the square root of the number of counts. As an example, for a resolution of 40 keV and 100 counts in a peak the statistical error of the excitation energy is 4 keV. Moreover, special care has to be taken to provide a very precise energy calibration and to eliminate systematical errors. For a detailed discussion, see Ref. [10]. As an example, the focal plane of a spectrometer is known to be in first order linear in momentum. But Grand Raiden has a deviation from linearity of about 100 keV in measurements of the (p,t) reaction at 200 MeV, that needs to be taken into account.

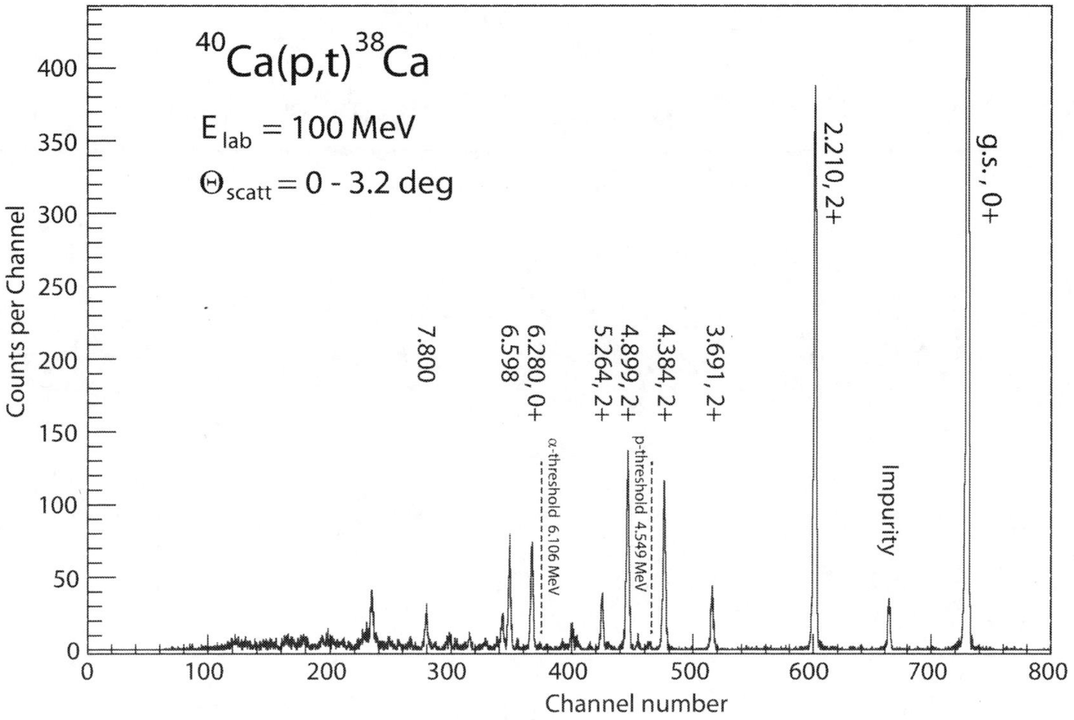

Figure 2. Measured ^{40}Ca(p,t)^{38}Ca spectrum. Peaks of levels are labeled by their excitation energies in MeV and their spin and parity where known.

4. Astrophysical studies

Network calculations of the rp-process give overall agreement with measured light curves of X-ray bursts. However, ever improving astronomical observations provide data that require more detailed information about the reaction paths of the rp-process. Detailed measurements of the light curves of X-ray bursts show ripples [15] and even double and triple peaks [16] separated by several seconds that are not explained by present network calculations or dynamical models.

The two main waiting-point nuclei, ^{14}O and ^{18}Ne, in the CNO cycles are by-passed by the ^{14}O$(\alpha,p)^{17}$F$(p,\gamma)^{18}$Ne$(\alpha,p)^{21}$Na$(p,\gamma)^{22}$Mg reaction sequence triggering the αp-process. The time-scale for the subsequent αp-process sequence ^{22}Mg$(\alpha,p)^{25}$Al$(p,\gamma)^{26}$Si$(\alpha,p)^{29}$P$(p,\gamma)^{30}$S(α,p) ^{33}Cl$(p,\gamma)^{34}$Ar$(\alpha,p)^{37}$K$(p,\gamma)^{38}$Ca and the switch over to the classical rp-process is determined by the associated reaction rates. If the (α,p) rates become low because of the increasing Coulomb barrier the reaction flow will be delayed because of the relatively slow β^+ decay of the waiting-point nuclei ^{22}Mg, ^{26}Si, ^{30}S, and ^{34}Ar. The (α,p) reaction rates are determined by the number of resonances, the resonance energies and α-partial widths of the α-unbound states within about 2 MeV above the α-thresholds of ^{30}S, ^{34}Ar, and ^{38}Ca, which are 9.342, 6.739 and 6.106 MeV, respectively. Similarly, the (p,γ) reaction rates are determined according to equation 1 by the number of resonances, the resonance energies and p-partial widths of the p-unbound states within about 2 MeV above the p-thresholds, which are 4.399, 4.663 and 4.549 MeV, respectively. To test this model for the waiting-point ^{34}Ar, we performed the ^{40}Ca$(p,t)^{38}$Ca reaction.

Journal of Physics: Conference Series **387** (2012) 012003 doi:10.1088/1742-6596/387/1/012003

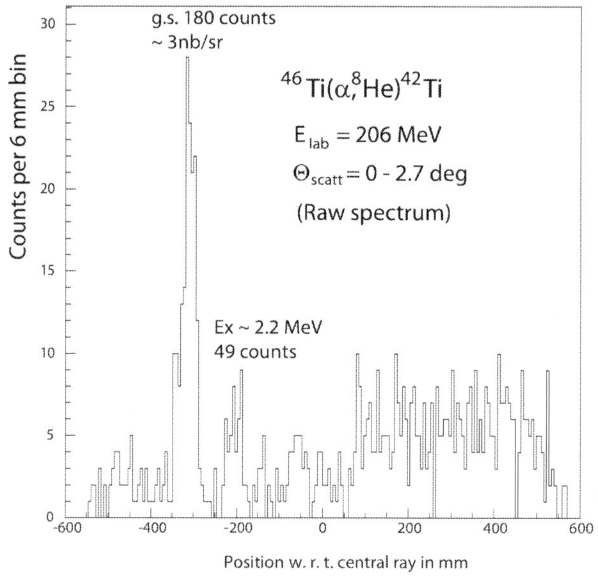

Figure 3. Measured ^{46}Ti(α,^{8}He)^{42}Ti spectrum. The peak due to the ground state and a level at around 2.2 MeV are indicated.

The (p,t) reaction has many advantages when measured with a spectrometer. The triton rigidity is more than a factor of 1.5 larger compared to the proton beam and other reaction products so that the beam can be stopped at the low momentum side of the dipole magnet and background from other reactions is small because they are also swept away from the focal plane. Figure 2 shows a sample spectrum, measured using the K600 spectrometer at iThemba LABS. The data analysis of this experiment is in progress.

Another astronomically interesting reaction is the ^{38}Ca(α,p)^{41}Sc, which is presently assumed to be the last αp-process in X-ray bursts based on shell-model calculations. The resonance energies are located above the α-threshold at an excitation energy of $E_x = 5.487$ MeV in the compound nucleus ^{42}Ti. These may be wrong by 100 - 200 keV as shown in a previous experiment [11]. Figure 3 shows the preliminary spectrum of the ^{46}Ti(α,^{8}He)^{42}Ti reaction in search for the resonances using Grand Raiden at RCNP, Osaka. These experiments are very difficult because of the low cross section of about 3 nb/sr for the ground state and about a factor of 10 smaller for excited states. Also, the (α,^{6}He) reaction has much larger yields which have to be discriminated against using the time-of-flight method and ΔE/E signal ratios from plastic scintillation detectors.

5. Outlook

We have shown that reaction studies of the nuclear structure with high-resolution spectrometers can provide valuable information to better understand the rp- and αp-processes. Great effort has been spent to measure the reaction cross sections of the (p,γ) and (α,γ) fusion reactions. This is also possible using electromagnetic recoil separators with very high mass resolution, by studying the reaction in inverse kinematics. This method has the advantage to be free of the beam related γ background, thus allowing to measure at lower energies closer to the corresponding temperature in the stellar medium. For this purpose the St. George recoil separator was designed and built at the University of Notre Dame to study the (α,γ) reaction using stable beams.

If a recoil separator is installed at a high-intensity radioactive beam facility as planned for

Journal of Physics: Conference Series **387** (2012) 012003 doi:10.1088/1742-6596/387/1/012003

FRIB [17], the reaction rates of selected reactions of the rp- and αp-processes that take place in the unstable proton-rich mass region can be measured. The project SECAR has been proposed and is under design for this purpose of studying the (p,γ) and (α,γ) fusion reactions initially at the present ReA3 facility with re-accelerated beams at MSU. When FRIB becomes available SECAR will reach its full potential with the higher intensity of radioactive beams.

Acknowledgements
This work would not have been possible without the collaboration of the researchers and the help of the cyclotron staff of RCNP and iThemba LABS. Their effort of providing a high quality beam and maintaining a facility with best performance is gratefully acknowledged. Special thanks go to John Greene of Argonne National Laboratory, whose continued supply of some of the most difficult targets were essential for these experiments.

References
[1] Hellier C 2001 *Cataclysmic Variable Stars* (Springer Press)
[2] Wallace R K and Woosley S E 1981 *Astrophys J Suppl* **45** 389
[3] Rolfs C E and Rodney W S 1988 *Cauldrons in the Cosmos* (University of Chicago Press, Chicago)
[4] Crowter R 2007 *M Sc thesis* (University of Surrey, Guildford, UK) unpublished
[5] Computer code NON-SMOKER ⟨http://nucastro.org/nonsmoker.html⟩
[6] Computer code TALYS ⟨http://www-astro.ulb.ac.be/Nucdata/Talys/pcapAL⟩
[7] Schatz H *et al.* 1998 *Phys Rep* **294** 167
[8] Azuma R E *et al.* 2010 *Phys Rev C* **81** 045805
[9] Berg G P A *et al.* 2003 *Nucl Phys A* **718** 608c
[10] Matic A *et al.* 2009 *Phys Rev C* **80** 055804
[11] Schatz H *et al.* 1997 *Phys Rev Let* **79** 3845
[12] Fujita Y *et al.* 1997 *Nucl Instr Meth B* **126** 274
[13] Wakasa T *et al.* 2002 *Nucl Instr Meth A* **482** 79
[14] Neveling R *et al.* 2011 *Nucl Instr Meth A* **654** 29
[15] Galloway D *et al.* 2004 *The Astrophys J* **601** 466
[16] Zhang G *et al.* 2009 *Mon Not R Astron Soc* **398** 8
[17] ⟨http://www.frib.msu.edu/⟩

The FrPNC Experiment, weak interaction studies in Francium at TRIUMF

E Gomez[1], S Aubin[2], R Collister[3], J A Behr[4], G Gwinner[3], L A Orozco[5], M R Pearson[4], M Tandecki[3], D Sheng[5], J Zhang[5]

[1] Institute of Physics, Universidad Autonoma de San Luis Potosi, San Luis Potosi, SLP 78290, Mexico

[2] Department of Physics, College of William and Mary, Williamsburg, VA 23187, USA

[3] Department of Physics and Astronomy, University of Manitoba, Winnipeg, Manitoba R3T2N2, Canada

[4] TRIUMF, Vancouver, British Columbia V6T 2A3, Canada

[5] Joint Quantum Institute, Department of Physics, University of Maryland, College Park, Maryland 20742, USA

E-mail: egomez@ifisica.uaslp.mx

Abstract. Francium is an excellent system to study the nuclear weak force due to its large nucleus and relatively simple atomic structure. The FrPNC experiment has a facility to produce cold trapped atomic francium samples for parity non-conservation studies. We are preparing to measure both the nuclear spin independent and dependent parts of the weak interaction in francium. The first one gives information about weak neutral currents at low energies, while the second one is sensitive to weak interactions between nucleons. We present the current status of the experiment.

1. Atomic Parity Non Conservation

Nuclear forces have been studied traditionally by bombarding the nucleus with a high energy projectile and analyzing the kinematics of the dispersed particles. A different possibility for the study of the weak force is by Parity Non Conservation (PNC) measurements in atoms. This technique is actually not that different from the previous one, since there is still a collision with the nucleus, though weak and continuous, but now the projectile is one of the electrons in the atom. The momentum of the electron in the atom is much smaller to what can be obtained in particle accelerators giving smaller weak interaction strength, but that is compensated in the case of atomic PNC by repeated collisions by the same electron that provides long integration times such that the signal can be extracted with high precision measurements. Atomic PNC measurements are useful for studying the nuclear weak force at low energies (around 1 MeV).

Atomic PNC measurements look for subtle deviations in the behavior of the electrons in the atom due to interactions with the nucleus. The interactions coming from the nuclear weak force can be separated from other interactions due to the fact that they violate parity. Since time reversal symmetry is preserved at the measured level, the experiments look for the existence of a parity forbidden transition that becomes allowed due to the weak force. The signal grows with a bigger nucleus since the electron has more nucleons to interact with. Moreover, a nucleus with higher Z has a higher electronic density at the nucleus and the electrons have higher velocity. The three effects combine to give a scaling of the (nuclear spin independent) signal faster than Z^3 [1] making heavy atoms the preferred ones. The extraction of weak interaction constants from the measurements rely on atomic structure calculations [2]. Francium is ideal for atomic PNC studies, since it has a large nucleus and a relatively simple atomic structure [3,4]. Furthermore, it features in enhanced signals for testing other fundamental symmetries, such as time reversal symmetry in electric dipole moment (EDM) measurements of the atom [5].

2. The FrPNC experiment

The FrPNC experiment is devoted to the study of weak interactions using francium atoms. We plan to send Fr ions produced in the ISAC facility at TRIUMF to a Y neutralizer foil. After sufficient accumulation of ions we rotate the foil and heat it to release neutral Fr atoms to a high efficiency laser cooling trap [6]. After cooling the atoms, a laser pulse pushes the sample down to a science vacuum chamber where we recapture the atoms in a second trap [7]. We expect to trap an isotopically pure sample of more than 10^6 atoms confined to a small volume (less than 0.1 mm^3) and at a very low temperature (around 100 μK). The ISAC facility can produce a selection of isotopes in the range of A=203-229. The availability of different isotopes is a particularly interesting feature of the experiment compared to other atomic PNC experiments giving us access to both the neutron deficient and neutron rich francium nuclei. PNC measurements in different isotopes are sensitive to new physics and will help determine neutron skins that will reduce PNC errors induced by them in traditional measurements [4,8]. The long interaction times available with laser cooled atoms allow for coherent measurements that give higher precision, while the absence of atomic motion and the small confinement volume suppress velocity and bias field uniformity related systematic errors.

A 500 MeV proton beam from TRIUMF collides with a uranium carbide target to produce Francium. The ions are extracted with 60 kV and the selected isotope is separated at ISAC. The demonstrated yields for the different isotopes produced reaches values up to 10^8 s^{-1} [9]. The 1% trapping efficiency [6] combined with the production yield gives 10^6 atoms, more than enough for PNC measurements [10]. We have constructed a radio frequency and electromagnetic interference shielded modular room to house the experiment (Fig. 1). The room has an attenuation of 100 dB at electromagnetic frequencies from 14 kHz to 10 GHz that are present in an accelerator environment and also strongly reduces acoustic noise. The temperature and humidity control is currently being installed and should be stable to 0.5°C and 5% relative humidity (at 45%) year round to maintain the lasers in a stable operating environment.

The weak interaction in atoms has a nuclear spin dependent and an independent part [11]. The nuclear spin independent part is the dominant one since it adds coherently over all nucleons. The magnitude of this interaction has been exquisitely measured in Cesium by quantifying the transition amplitude between the $6S_{1/2}$ to $7S_{1/2}$ levels [12]. The measurement together with atomic structure calculations give the value of the weak charge (Q_W). The weak charge is related to the Weinberg's angle and atomic PNC gives a value of $\sin^2\theta_W$=0.2382(11) [2] at low energy (below 1 MeV). Francium is heavier than Cesium and the PNC effect is expected to be 18 times larger. The Cesium experiment uses an atomic beam that has the advantage of having a large atomic flux. The number of

Journal of Physics: Conference Series **387** (2012) 012004 doi:10.1088/1742-6596/387/1/012004

francium atoms will be smaller, but since they remain trapped, the number of atoms in the interaction region at any given time is similar to the experiments with Cs [13].

Fig. 1. Shielded room housing the FrPNC experiment in the ISAC hall. The metallic room is located below the blue platform.

The planned measurement of the weak charge follows the technique pioneered by Wieman and coworkers in Cesium [12]. A laser induces a transition between the $7S_{1/2}$ to $8S_{1/2}$ levels in Fr. The excited atoms decay back to either hyperfine level of the ground state. By initially pumping all the atoms into the lower ground hyperfine state, we can make use of a shelving technique to put the atoms that were excited into the upper hyperfine ground state for high efficiency optical detection using a cycling transition [14]. The interpretation of the results requires a good understanding of the atomic structure of Fr, and in that direction extensive spectroscopic studies have been performed over the years [11]. Three important quantities remain to be measured precisely for the eventual extraction of the weak charge: the M1 transition amplitude between the levels of interest [15] and the scalar and tensor polarizabilities (α and β). The measurement of the polarizabilities involves Stark induced transitions between different Zeeman levels. The M1 transition can be measured via an $E1_{Stark}$-M1 interference. These measurements serve as preparation steps for the more challenging PNC experiment since the method is very similar but with stronger transition amplitudes.

The spin dependent nuclear weak interaction is also of interest since it produces the nuclear anapole moment, which can be related to the weak nucleon coupling constants. The anapole moment has been measured in Cesium [12] and a number of other atomic and molecular PNC experiments are searching for anapole moments as well [16]. The expected signal in Fr is not only 11 times larger than in Cs, but the availability of different isotopes opens the possibility of separating the contributions to the anapole moment coming from valence protons or neutrons. The constraints produced by each one are no longer parallel, giving a crossing region for the weak nucleon coupling constants from a single experiment [13]. A complementary measurement in rubidium is an intriguing option since there are two stable isotopes available that have a valence proton in a different orbital that should produce an anapole signal of opposite sign [7]. The anapole moment can be measured by inducing E1 transitions between hyperfine levels in the presence of a static magnetic field [10]. A coherent interference with an M1 transition gives a signal to noise ratio larger than 1 (in 1s), a very large ratio for a PNC measurement. The extraction of the anapole moment from the measurement requires a good understanding of the nuclear structure involved [4,17].

3. Recent progress

There has been some preliminary work towards an anapole moment measurement using Rb at the University of Maryland. Out of the requirements we demonstrated: long coherence time M1 transitions, the sensitivity of the interference method, a blue detuned trap and a requirement for a pure transverse mode of the microwave cavity. The blue detuned trap uses two perpendicular acousto-optic modulators fed by two phase locked function generators (SRS DG345) to dynamically change the shape of the trap [18]. The measurements using the M1 clock transition give a 180 ms coherence time, which is much larger than the one we obtain using optical Raman transitions [19]. Furthermore, the microwave M1 transitions are simpler to use than the Raman transitions and give less stringent constraints in the mechanical stability of optical elements. To test the interference method we use two M1 fields, one represents the interfering M1 transition and the other one has a Rabi frequency 100 times smaller and simulates the E1 anapole allowed transition. We can resolve the second field with a signal to noise ratio of 20 [19]. We quantified the differential AC Stark shift of the transition in the dipole trap and found a scaling inversely proportional to the detuning. We are designing and constructing a second generation microwave Fabry-Perot cavity, following a quasi-optics solution. The mirrors feature imprinted patterns to control the reflectivity and polarization of the circulating microwaves and the mode will be controlled using the appropriate microwave horns. The mirror design benefitted from calculations using the HFSS electromagnetic field simulation software [20]. Further progress will occur at TRIUMF where we have now a working Rb trap (Fig. 2).

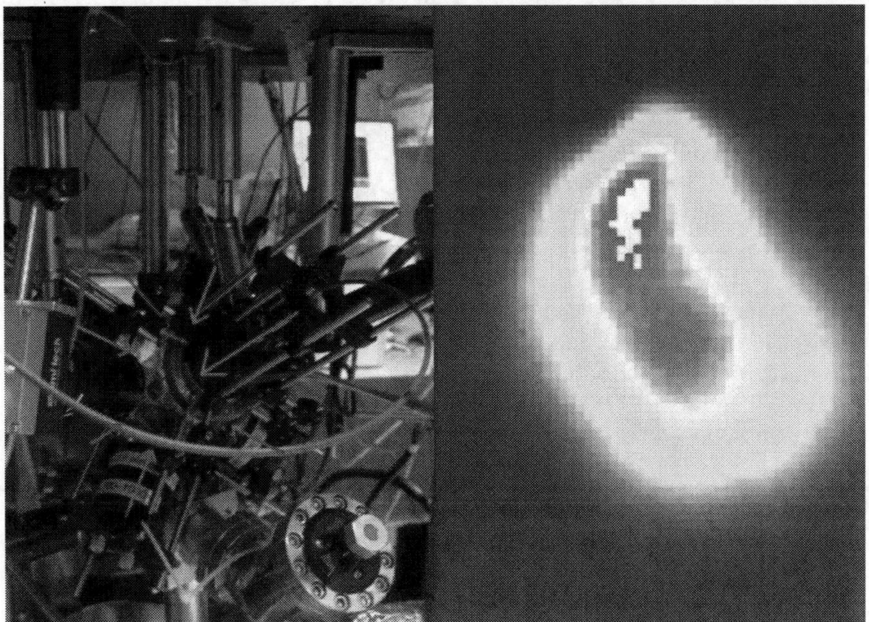

Fig. 2. Optical fiber-based laser cooling trap at TRIUMF (left, with red arrows indicating the trapping laser beams) and fluorescence image of the trapped Rb atoms (right).

In summary, the FrPNC collaboration has now moved to TRIUMF where we are building a facility for obtaining cold and trapped francium atoms. We plan to use the francium for atomic PNC measurements to probe and study the weak nuclear force. A measurement of the nuclear spin-independent part can be used to determine the weak charge and the weak-mixing Weinberg angle at low energies, while a measurement of the nuclear spin-dependent PNC and the anapole moment will help constrain the weak nucleon coupling constants.

Work supported by DOE, NSF, NSERC, CONACYT and TRIUMF

References

[1] Bouchiat M A and Bouchiat C 1997 *Rep. Prog. Phys.* **60** 1351

[2] Porsev S G and Beloy K and Derevianko A 2010 *Phys. Rev.* D **82** 036008

[3] Ramsey-Musolf M J and Page S A 2006 *Annu. Rev. Nucl. Part. Sci.* **56** 1

[4] Sil T and Centelles M and Viñas X and Piekarewicz J 2005 *Phys. Rev.* C **71** 045502

[5] Mukherjee D and Sahoo B K and Nataraj H S and Das B P 2009 *J. Phys. Chem.* A **113** 12549

[6] Aubin S and Gomez E and Orozco L A and Sprouse G D 2003 *Rev. Sci. Instr.* **74** 4342

[7] Sheng D and Orozco L A and Gomez E 2010 *J. Phys.* B **43** 074004

[8] Brown B A and Derevianko A and Flambaum V V 2009 *Phys. Rev.* C **79** 035501

[9] http://www.triumf.info/facility/research_fac/yield.php?element_name=Fr

[10] Gomez E and Aubin S and Sprouse G D and Orozco L A and DeMille D P 2007 *Phys. Rev.* A **75** 033418

[11] Gomez E and Orozco L A and Sprouse G D 2006 *Rep. Prog. Phys.* **69** 79

[12] Wood C S and Bennett S C and Cho D and Masterson B P and Roberts J L and Tanner C E and Wieman C E 1997 *Science* **275** 1759

[13] Behr J A and Gwinner G 2009 *J. Phys.* G: *Nucl. Part. Phys.* **36** 033101

[14] Wood C S and Bennet S C and Roberts J L and Cho D and Wieman C E 1999 *Can. J. Phys.* **77** 7; Rowe M A and Freedman S J and Fujikawa B K and Gwinner G and Shang S Q and Vetter P A 1999 *Phys. Rev.* A **59** 1869

[15] Savukov I M and Derevianko A and Berry H G and Johnson W R 1999 *Phys. Rev. Lett.* **83** 2914

[16] Ferro F and Surzhykov A and Stöhlker T 2011 *Phys. Rev.* A **83** 052518; DeMille D and Cahn S B and Murphree D and Rahmlow D A and Kozlov M G 2010 *Phys. Rev. Lett.* **100** 023003; Mandal P and Mukherjee M 2010 *Phys. Rev.* A **82** 050101(R)

[17] Kanjilal D and Saha S and Bhattacharya S and Goswami A and Kshetri R and Raut R and Muralithar S and Singh R P and Mukherjee G and Mukherjee B 2011 *Phys. Rev.* C **84** 064321

[18] Sheng D and Perez Galvan A and Zhang J and Orozco L A "Atoms in a blue detuned dipole trap: studies of coherence and dynamics for precision measurements in the classical and quantum regimes." To be submitted

[19] Sheng D and Zhang J and Orozco L A "Sensitivity test of blue-detuned dipole traps designed for parity non-conservation measurements in Fr." Submitted

[20] Zhang J and Hood J D and Orozco L A "Microwave Fabry-Perot open resonator for francium spectroscopy." To be submitted

Journal of Physics: Conference Series **387** (2012) 012005 doi:10.1088/1742-6596/387/1/012005

Systematics of Low Energy Collective States in neutron-rich Cd Isotopes

J C Batchelder[1], N T Brewer[2], C J Gross[3], R Grzywacz[3,4], J H Hamilton[2], M Karny[3,5], A Kuzniak[5], M F Madurga[4], A J Mendez[3], S H Liu[1], K Miernik[3], D Miller[4], S W Padgett[4], S V Paulauskas[4], K P Rykaczewski[3], A V Ramayya[2], D W Stracener[3], M Wolinska-Cichocka[3,5], and E F Zganjar[6]

[1] UNIRIB/ Oak Ridge Associated Universities, Oak Ridge TN 37831, USA
[2] Vanderbilt University, Nashville TN 37235, USA
[3] Physics Division, Oak Ridge National Laboratory, Oak Ridge TN 37931, USA
[4] University of Tennessee, Knoxville TN 37996, USA
[5] Oak Ridge Associated Universities, Oak Ridge TN 37831, USA
[6] Louisiana State University, Baton Rouge, LA 70803, USA

E-mail: batchelderjc@ornl.gov

Abstract. It has been shown that there are significant deviations from the expected U(5) dynamical symmetry for ^{110,112,114,116}Cd. However, there is very significant mixing with intruder states in this region. In this paper, we investigated states in the heavier ^{120,124,126}Cd populated via beta decay. These nuclei exhibit similar patterns to the lighter Cd isotopes even though the intruder states are much higher in energy.

1. Introduction

Up until recently, the neutron-rich even-even Cd isotopes $^{110\text{-}114}$Cd are often cited as textbook examples [1,2] of vibrational nuclei and the best examples of U(5) symmetry. Nuclei that are close to the limits of U(5) symmetry would be expected to exhibit harmonically spaced multiphonon states. In the case of the Cd isotopes. This picture is complicated by low-lying intruder states (caused by the elevation of 2-protons across the Z = 50 shell gap) which at the proton mid-shell are in the same energy region as the 2-phonon states. As such, the lowest lying levels in the neutron-rich even-even Cd isotopes have energies that resemble an anharmonic vibrator coupled to 2-proton intruder states.

The experimental signature of these N-phonon states is a strong preference for decay of a N-phonon state to the (N-1)-phonon states over other lower energy states. The contribution of the intruder states to the B(E2) values of the transitions from the N-phonon states can be accounted for by calculations in the IBM-2 framework [3,4].

Previous studies have reported candidates for the 3-phonon states in ^{110}Cd [5], ^{112}Cd [3], ^{116}Cd [6], ^{118}Cd [7] and ^{120}Cd [8]. However, this simple picture was shown to be at odds with the experimental data for the cases of ^{110,112,114}Cd [4,9] and ^{116}Cd [10] where it was demonstrated that while the 6^+, 4^+ and 3^+ states decay as expected, the previously proposed 0^+ and 2^+ candidates for the three-phonon quintuplet are not in agreement with IBM-2 calculations. The disagreements with calculations can not be due to experiments simply missing the transitions, as the nuclei ^{110,112,114,116}Cd have all been studied by a variety of methods with very high statistics.

Journal of Physics: Conference Series **387** (2012) 012005 doi:10.1088/1742-6596/387/1/012005

This discrepancy in the decay of the suggested three phonon 0^+ and 2^+ states with the U(5) description is a consistent feature in all the well-studied even-even neutron-rich Cd nuclei. Figure 1 shows the decays of the 0^+ and 2^+ members of the reported 3-phonon quintuplet for 112Cd [11], ^{114}Cd [12], ^{116}Cd [10], and ^{118}Cd [8]. None of these nuclei decay as expected in a vibrational picture.

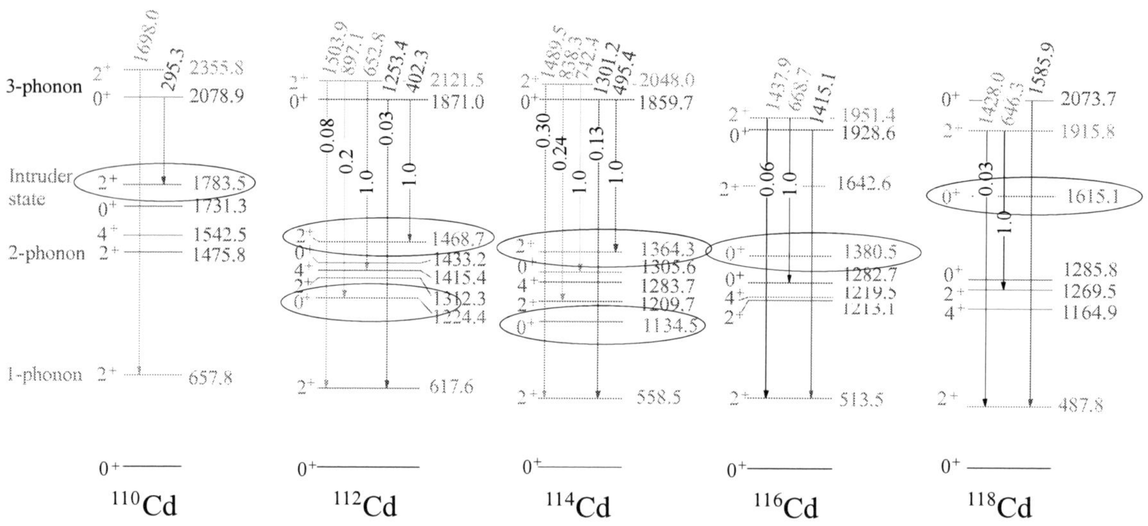

Figure 1. Comparison of the decays of the 0^+ and 2^+ members of the three phonon levels in ^{112,114,116,118}Cd. Relative B(E2)s are shown. The intruder states are circled for clarity.

It is well established that as one moves away from the neutron mid-shell the energy of the intruder states would be expected to rise. If the explanation for the discrepancy between observed decays of the candidate 3-phonon states and those expected from IBM-2 calculations at least partially arises from mixing with intruder states, the picture should become more clear the further from neutron mid-shell. A systematic study of these nuclei is therefore needed to fully understand what is happening in the neutron rich Cd nuclei. In particular, how well does this relatively simple model describe the low-lying levels in the neutron-rich even-even Cd nuclei? With this as our motivation, we have remeasured the low energy states in ^{120,122,124,126}Cd via the beta decay of ^{120,122,124,126}Ag.

2. Experimental Method

Silver-120,122,124,126 were produced via the proton-induced (40 MeV) fission of ^{238}U at the Holifield Radioactive Ion Beam Facility (HRIBF) in a plasma ion source and then transported to a moving tape collector. At the HRIBF, there are two locations available for these types of experiments. The first is the On-Line Test Facility (OLTF) (formally called UNISOR) which is connected to an online mass separator ($\Delta m/m \sim 2000$). The 40 MeV protons delivered to this facility are from the 25MV tandem and have an administrative beam current limit of 50 nA. The second involves 50 MeV protons of up to 20 μA from the ORIC cyclotron. The beam is then separated with a high resolution ($\Delta m/m \sim 10{,}000$) magnet and delivered to the LeRIBSS (Low-Energy Radioactive Ion Beam Spectroscopy Station) detector array. The experiment presented herein for the decay of ^{120}Ag was performed at the OLTF separator, and the experiments on ^{122,124,126}Ag are from the ORIC/LeRIBSS setup. Results are presented from the ^{120,124,126}Ag decays, while results from ^{122}Ag will be presented in a future paper [13].

The detector setup at the OLTF is the CARDS array (Clover Array for Radioactive Decay Spectroscopy), which consisted of three segmented-clover Ge detectors and a Si conversion-electron spectrometer. The clover detectors were mounted in a close geometry surrounding the tape forming three sides of a square ~10 cm per side. The data acquisition system utilized in this work was a digital spectroscopy system based on DGF-4C modules (produced by X-ray Instrumentation Associates). These modules incorporate 40 MHz flash ADC's, and serve as a replacement for amplifiers, discriminators, conventional ADC's and TDC's. There were two sets of data taken. The first set of data was taken in "take-away" mode where the beam is collected in the center of the detector array, and the tape is moved every three seconds to remove longer-lived contaminants and daughter products. This setting was the bulk of the run with a total of ~ 75 hours of data taken. In the second set of data (of ~ 5 hours), the beam was collected ~ 30 cm in front of the Ge-detectors and then moved into the center of the array. The tape was moved every three seconds with a tape transport time of ~800 ms. This setting was used primarily to measure half-lives and resulted in a very reduced rate of ^{120m}Ag. More details on the experimental setup may be found in Refs. [14,15,16].

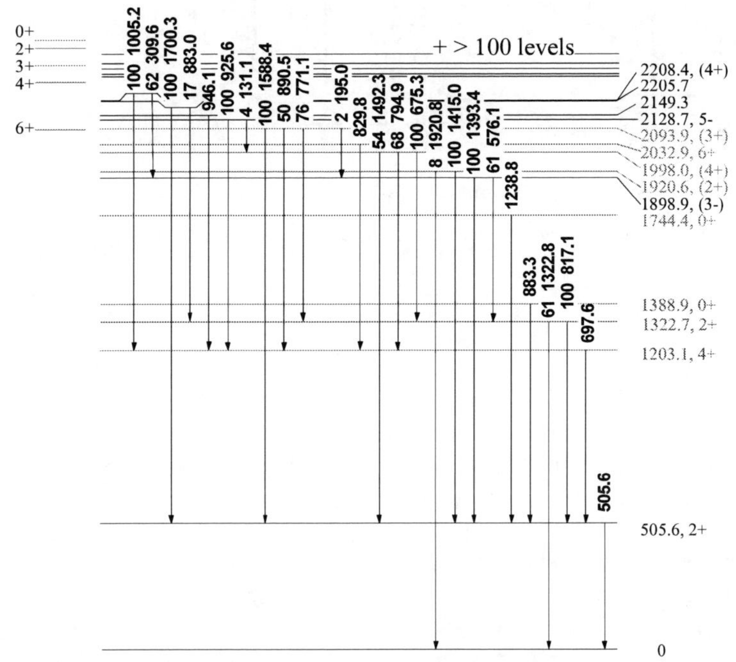

Figure 2. Low energy levels in ^{120}Cd populated in the beta decay of ^{120m,gs}Ag.

At LeRIBSS the detector setup was similar, except there were four segmented clover Ge detectors, and two plastic scintillators in the CARDS setup. The electronics used were the PIXIE-16 modules [17] which are a newer, faster version of the DGFs. Due to the very short half-lives involved, the activity was collected in the center of the array. In order to measure half-lives, the beam was pulsed via electrostatic plates. The total tape cycle used for these measurements is 3 seconds, with the beam on for the first 2 seconds, and then turned off for one second to allow the half-life measurement.
Overall, very high statistics (more than a factor of 50 larger than previous in all cases) were collected for these nuclei compared to previously published results. The number of events observed in the $2^+_1 \rightarrow 0^+$ transition were ^{120}Ag: 1e8, ^{122}Ag: 1e6, ^{124}Ag: 6e5, and ^{126}Ag: 1e5.

3. Experimental Results

3.1. ^{120}Ag

The beta decay of ^{120}Ag was first reported by Fogelberg *et al.*, [18]. In that work, 6 γ-rays from 5 levels in ^{120}Cd were measured from the decay of two isomers of ^{120}Ag. The most recently published work by Wang *et al.*, [8] increased this to 62 γ's depopulating 26 levels. Four of these levels 1899.0 (3^+), 1920.5 (2^+), 1997.9 (4^+), and 2032.8 (6^+), were assigned as candidates for three-phonon states in that work. Of these levels, the 6^+ and 4^+ decay as expected with a strong preference for decay to the 2-phonon states, but the 2^+ and 3^+ states do not. The assigned (3^+) state at 1899.0 keV decays only to the 2^+ 2-phonon state but not the expected 4^+ 2-phonon state. Previous experimental work based on ^{124}Sn(d,^{6}Li) had assigned this state as 3^- [19]. The 2^+ identified as a 2-phonon state in that work decays only to the 2^+ 1-phonon state. No candidates for a 3-phonon 0^+ state were reported.

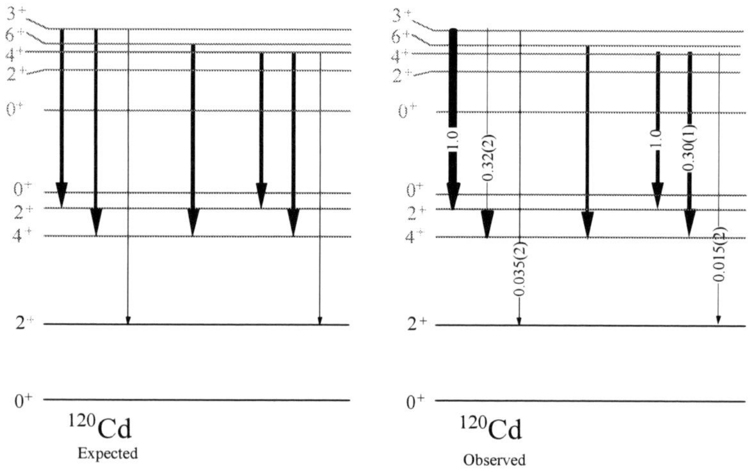

Figure 3. Comparison of expected versus observed decays of the 3^+, 6^+, and 4^+ "3-phonon states" in ^{120}Cd.

In the current work, a grouping of nine states (1898.9, 1920.6, 1998.0, 2032.9, 2093.9, 2128.7, 2149.2(new), 2205.7(new), and 2208.4 keV) between 1899 and 2208 keV were observed (see Figure 2). Of these, this work supports the J^π assignments from previous work for 1920.6 (2^+), 1998.0 (4^+), 2032.9 (6^+), 2128.7 (5^-), and 2208.4 (2^+). The state at 2093.9 was assigned 2^+ in Ref [20] based on a reported transition to the 0^+ ground state. No evidence of this transition is present in the current work either in singles or coincidence with the gammas feeding this state. This state feeds the levels at 505.6 (2^+), 1203.1 (4^+), and 1322.7 (2^+). We therefore assign a J^π of 3^+ to this state. We also adopt the 3^- value for the 1898.9 keV state from Ref [19]. The two new states at 2149.2 and 2205.7 keV likely have J^π of 2^+ or 4^+, however, neither decays in a manner consistent with the systematics of the lighter Cd "3-phonon states".

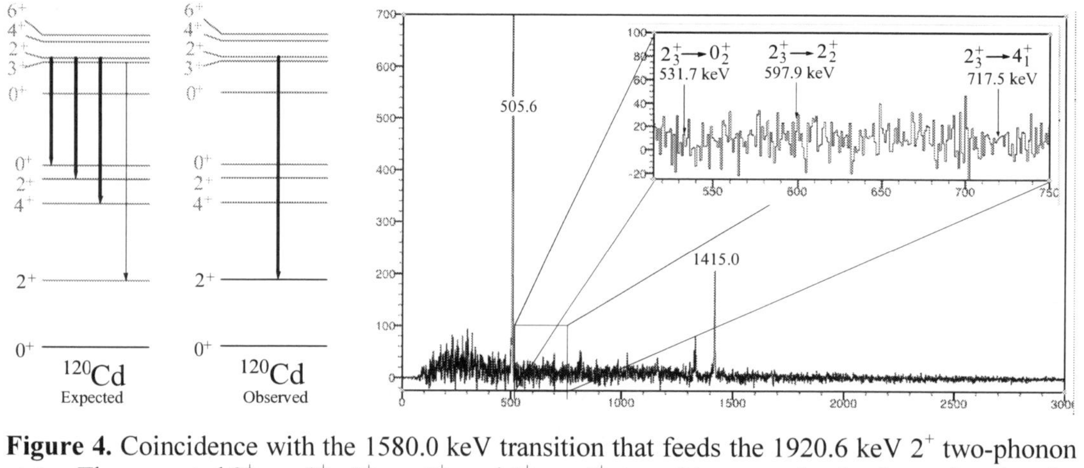

Figure 4. Coincidence with the 1580.0 keV transition that feeds the 1920.6 keV 2^+ two-phonon state. The expected $2^+_3 \rightarrow 0^+_2$, $2^+_3 \rightarrow 2^+_2$ and $2^+_3 \rightarrow 4^+_1$ transitions are clearly absent (see inset).

We assign the states at 1920.6 (2^+), 1998.0 (4^+), 2032.9 (6^+), and 2093.9 (3^+) as "3-phonon states" in ^{120}Cd. No candidates for the 0^+ member of the quintiplet were observed in the data. The 3^+, 4^+, and 6^+ states decay in a manner consistent with that of a 3-phonon state. Figure 3 shows a comparison of experimental relative B(E2)s versus the expected decay of 3^+, 6^+ and 4^+ multiphonon states. All three of these states decay as expected.

The 2^+ state is only observed to decay to the 1-phonon 2^+ state. A 2^+ member of the three-phonon quintuplet would be expected to strongly decay to all three two-phonon states (4^+, 2^+, 0^+) and weakly to the 2^+ one-phonon state. Figure 4 show the spectrum coincident with a 1580.0 keV transition that feeds the state. In this spectrum, the expected $2^+_3 \rightarrow 0^+_2$, $2^+_3 \rightarrow 2^+_2$ and $2^+_3 \rightarrow 4^+_1$ transitions are clearly absent. In these data, no suitable candidates for a 0^+ state in this energy region were observed.

3.2. ^{124}Ag

Previous work (via the beta decay of ^{124}Ag [21], and a ^{238}U(α,fγγγ) study [22] resulted in the discovery of a total of 16 states and 26 gammas (with spins up to 12^+). Of these, positive parity states at 613 (2^+_1), 1385 (4^+_1), 1428 (2^+_2), 1915 (3^+, 4^+), 1924 (4^+, 3^+), 1978 (2^+), and 2139 (6^+) are in the energy range that multi-phonon states would be expected. Notably absent are any states assigned as 0^+ (either intruder or phonon state.

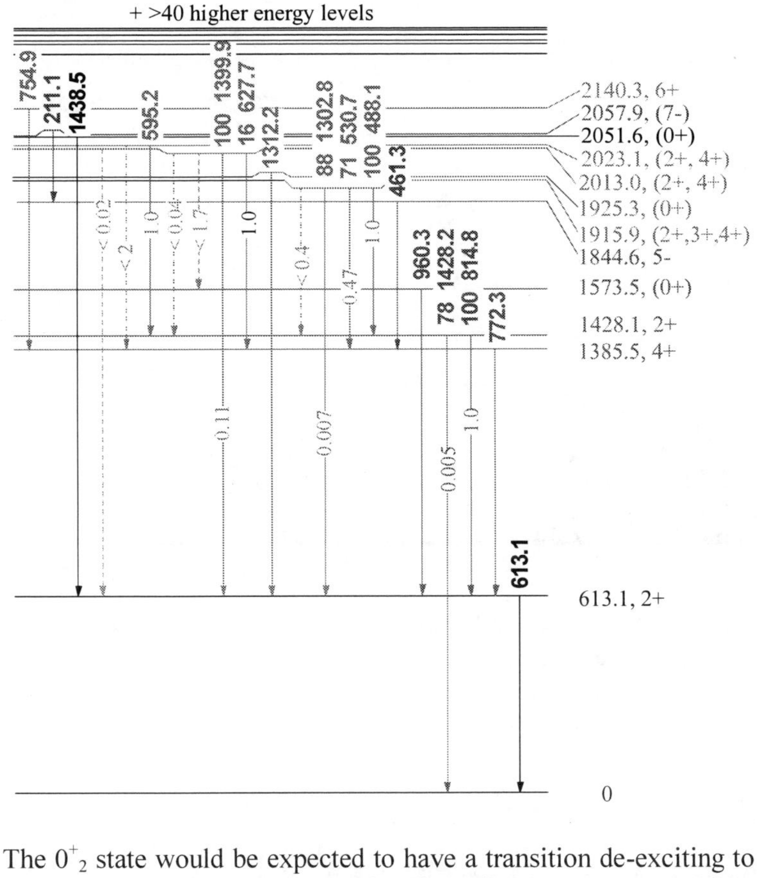

Figure 5. Low lying levels in ^{124}Cd showing a grouping of states between 1900 keV and 2150 keV that are possible three-phonon states. Relative B(E2)s are shown.

The 0^+_2 state would be expected to have a transition de-exciting to the 2^+_1 state with an energy ~700-900 keV. In addition, this state in all the lighter even mass Cd isotopes is only weakly fed (if at all) from above. Only one peak (960.3 keV) close to the expected energy is coincident with the 2^+_1 613.5 keV state and nothing else. We therefore assign this state at 1573.5 as (0^+_2). Many other new levels have been added to the decay scheme from this work, including 8 levels grouped in the energy region where a three phonon quintuplet would be expected (see Figure 5). The two levels at 1844.6 (5^-) and

2057.9 keV (7^-) were previously assigned as members of a negative parity band [22]. Our interpretation in the context of possible multiphonon states of the other levels are detailed below.

3.2.1. 1915.9 keV level

This level shows an enhanced decay to the 2^+_2 and 4^+_1 levels, and a weak decay to 2^+_1 level. No feeding into this state was observed from higher energy states. The decay of this state is similar to the decays of the 3^+_1 states in ^{116,118,120}Cd. We therefore assign this state as 3^+_1, however 2^+ and 4^+ are not ruled out.

3.2.2. 1925.3 keV level and 2051.6 keV level

The 1925.3 keV level is only observed to decay to the 2^+_1 level. No evidence of a 1925.5 keV gamma is present in the data. We have observed 2 levels feeding it from above. The two levels are at 2526.7 keV (which also decays to the 2^+_1 state), and 2763.0 keV which only decays to this state. The 2051.7 keV only decays to the 2^+_1 level. We assign both these levels as 0^+. The energies of the 0^+ intruder states in ^{118,120,122}Cd are 1615, 1744, and 1992 keV respectively. Cadmium-124 is even further from mid-shell so the energy of its intruder state will increase. Therefore, we assign the 2051.6 keV state as the 2p-2h intruder state.

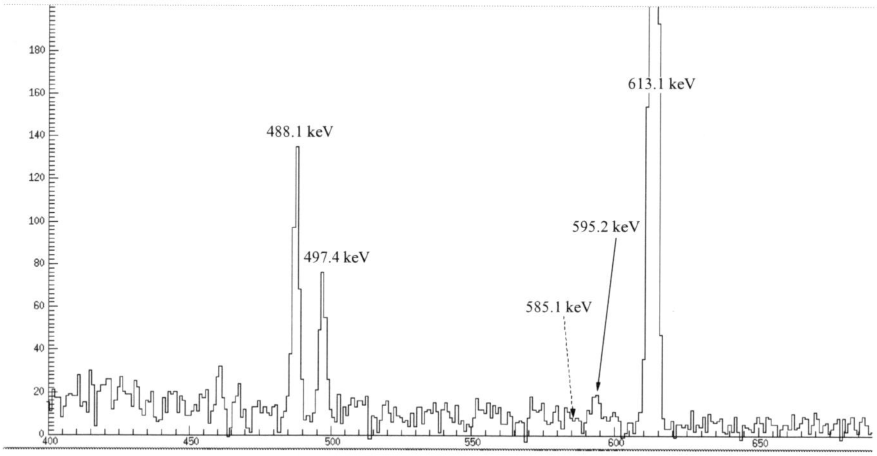

Figure 6. Gammas coincident with the 814.8 keV transition.

3.2.3. 2013.0 keV level

The 2013.0 keV level has an enhanced decay to the 4^+_1 level, and a weak decay to the 2^+_1 level, with no observed feeding from above. An expected 585.1 keV transition to the 2^+_2 level is not observed. Figure 6 shows a portion of the spectrum created by gating on the 814.8 keV $2^+_2 \rightarrow 2^+_1$ transition. The 585 keV gamma is clearly absent. The character of this state is not clear, but can be assigned as either 4^+ or 2^+.

3.2.4. 2023.1 keV level

The only observed decay from this state is to the 2^+_2 level. This state is only weakly populated in the beta decay of ^{124}Ag, and no feeding from above was observed. The decay of this state is similar to the lighter Cd "3-phonon" 4^+_1 states. We therefore assign this state as (4^+).

3.2.5. 2140.3 keV level

This level is only observed to decay to the 4^+_1 level. It is fed from above by the known (7^-) 2383.7 keV and 2675.2 keV (8^+) states [22]. We therefore accept the previous assignment of 6^+ for this state.

Overall, we have found five states that could be considered as candidates for three-phonon states. The decay of these states is very similar to the lighter Cd isotopes. The 6^+ (2140.3 keV), 4^+ (2023.1 keV) and 3^+ (1916.1 keV) members decay in a manner consistent with multi-phonon states, while the 0+ (1925.3 keV) and likely 2^+ member at 2013.0 do not.

3.3. ^{126}Ag

Previous studies on ^{126}Cd reported a total of 5 levels with 6 gammas via the beta decay of ^{126}Ag [21], and decay of a μs (12^+) isomer of ^{126}Cd [23]. The reported levels form what appears to be a ground state band and a negative parity band (with spins up to 9^-). These studies did not observe the 2^+_2 state. In this work, we report the observation of a state at 1579.4 keV that de-excites by a 927.3 gamma transition to the 2^+_1 and a 1579.4 keV gamma to the ground state. We assign this level as the 2^+_1 state based on the systematics of lighter Cd isotopes and predictions by the shell model code OXBASH [21]. The previous beta decay study did not observe this transition because it was masked by contamination from a 928.3 keV gamma-ray from the decay of ^{126}Sb [24].

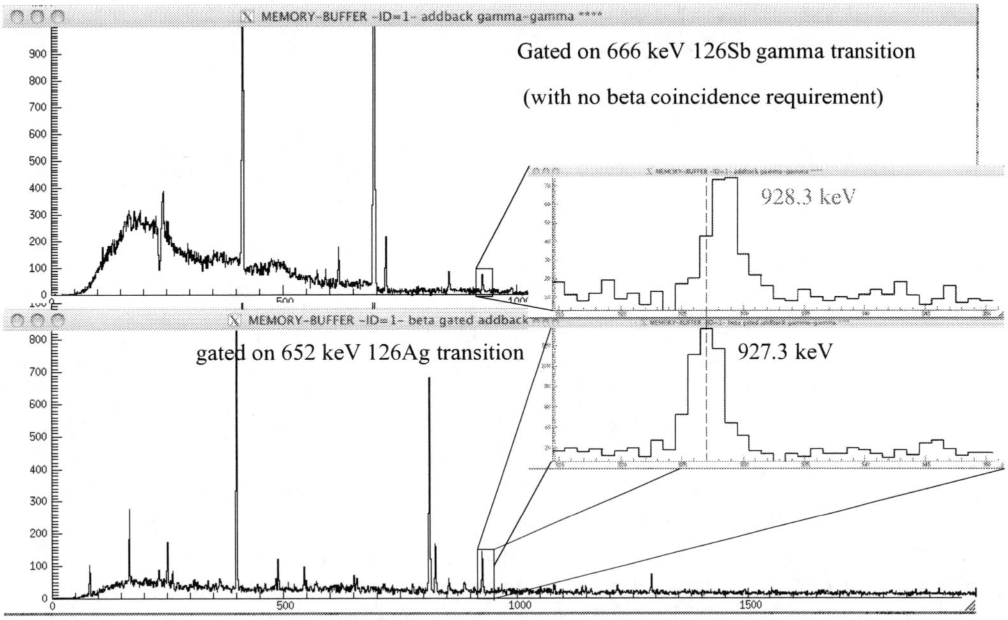

Figure 7. Gamma spectra created by gating on the (a) 666.4 keV $2^+_1 \rightarrow 0+$ from the ^{126}Sb beta decay, and (b) 652.0 keV $2^+_1 \rightarrow 0^+$ gamma transition from the beta decay of ^{126}Ag.

The present study also contained a small amount of contamination from ^{126}Sb. Figure 7a (*top panel*) shows the gammas coincident with the 666.4 keV $2^+_1 \rightarrow 0^+$ transition in ^{126}Te populated via the beta decay of ^{126}Sb. In this spectrum, a 928.3 keV gamma is present. In Figure 7b (*bottom panel*) gammas coincident with the 652.0 keV $2^+_1 \rightarrow 0^+$ transition in ^{126}Cd are presented, which clearly shows a peak at 927.3 keV peak -1 keV different from the ^{126}Sb contaminant peak (see inset).

Figure 8 shows the decay scheme below 2450 keV. Although there are a few new levels observed in the energy region of interest, they are spread out in energy, and are not good candidates for 3-phonon levels. Unlike the lighter even-mass Ag isotopes, the beta decay of ^{126}Ag is dominated by a large amount of feeding to negative parity states. Analog states to the ones discussed above for ^{120,124}Cd may be present in this nucleus, but are not observed in this work due to the small feedings of these states via beta decay. In order to make a conclusion about whether or not these states exist, either a lot more statistics is needed, or the states must be populated by another method.

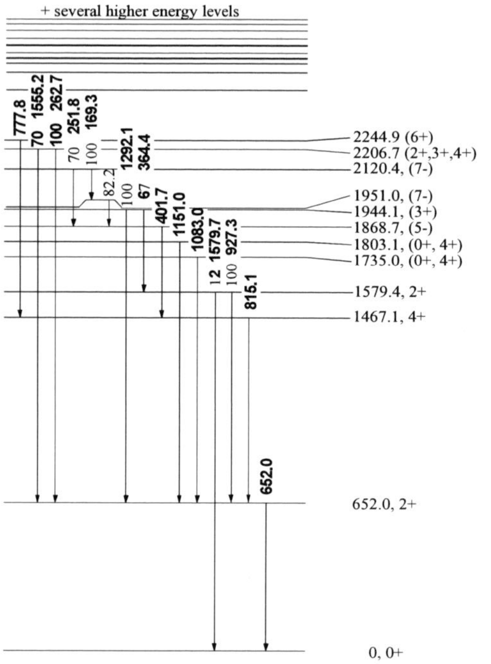

Figure 8. Partial level scheme of ^{126}Cd below 2450 keV populated by the beta decay of ^{126}Ag.

4. Conclusions

The decay pattern of ^{120,124}Cd are similar to lighter Cd isotopes despite the higher energy of the intruder states. Cadmium-120 and ^{124}Cd show a similar pattern as the lighter Cd isotopes, with no good candidates for 2^+ and 0^+ 3-phonon states. The current situation for ^{126}Cd is unclear as candidate 3-phonon states are absent in this work. Whether this is due to the states not being populated in beta decay or are simply absent cannot be determined.

Are the Cd nuclei really vibrational? Overall, the multi-phonon vibrational approach fails to explain the low-energy structure of these Cd isotopes. We propose an alternate way to describe the low-lying states in Cd by placing them in independent band structures rather than as N-phonon states. In all of these cases there is a ground state band, a quasi-gamma band, and an intruder band. This is illustrated in Figures 9-12 for ^{116}Cd [10], ^{118}Cd [8], ^{120}Cd, and ^{124}Cd. This interpretation is more consistent with the experimental data especially for the 0^+_4 and 2^+_4 states, than the three-phonon picture, as the decay pattern from these levels is what would be expected from the decay of independent bands.

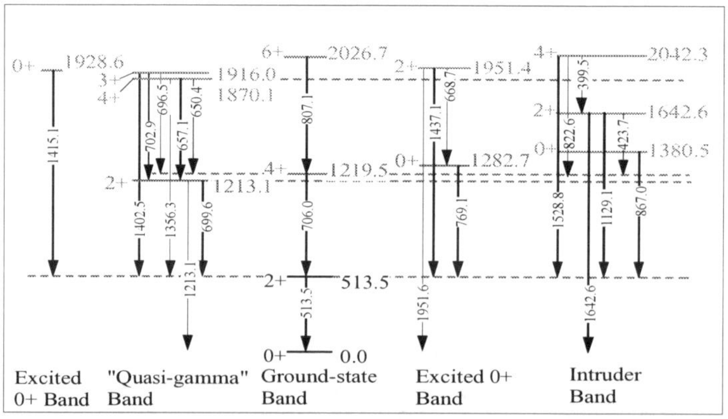

Figure 9. Low energy levels in ^{116}Cd drawn as band structures.

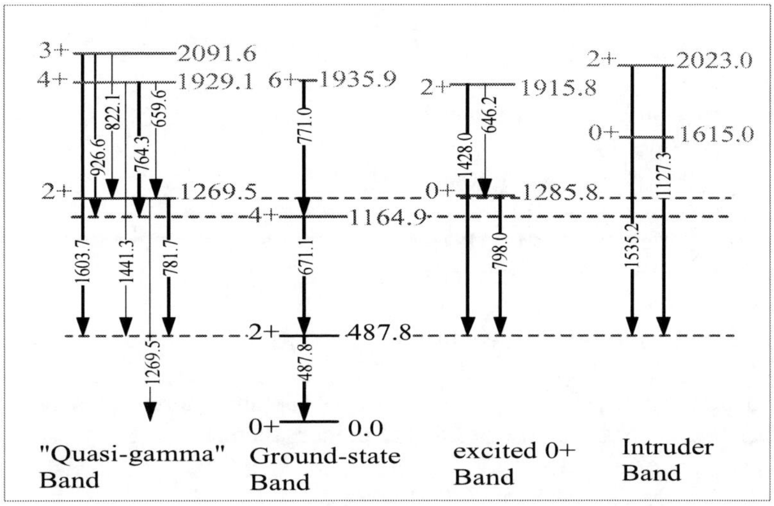

Figure 10. Low energy levels in ^{118}Cd drawn as band structures.

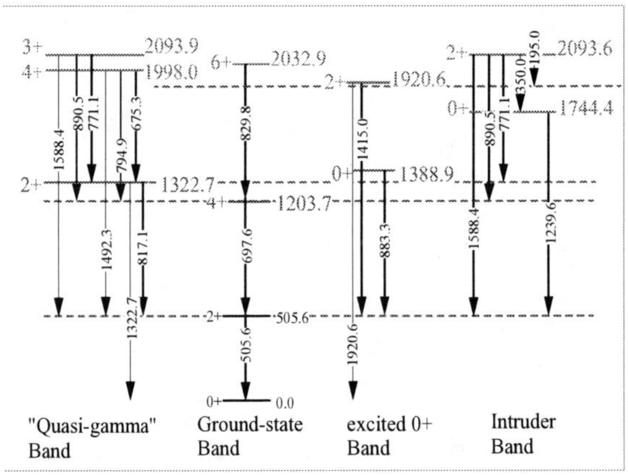

Figure 11. Low energy levels in ^{120}Cd drawn as band structures.

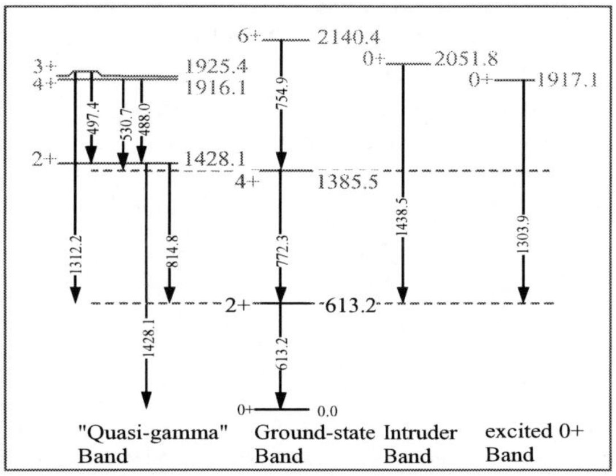

Figure 12. Low energy levels in ^{124}Cd drawn as band structures.

In this work, we have shown that the deviations from U(5) dynamical symmetry for the heavy even-mass Cd isotopes continue to higher masses even though the intruder states are further away in energy. While these states do not appear to be multiphonon states, their exact nature remains unclear. Future studies using other experimental methods (g-factors, etc.) will be needed to better understand the nature of these states.

Acknowledgments
This work has been supported by the U. S. Department of Energy under contracts DE-AC05-76OR00033 (UNIRIB), DE-FG05-88ER40407 (Vanderbilt University), DOE-AC05-00OR22725 (ORNL), DE-FG02-96ER40983 (University of Tennessee) and DE-FG02-96ER40978 (Louisiana State University).

Journal of Physics: Conference Series **387** (2012) 012005 doi:10.1088/1742-6596/387/1/012005

References

[1] Casten R F 2000 *Nuclear Structure from a Simple Perspective*, 2nd Ed. (Oxford University Press) pp 191
[2] Kern J, Garrett P E, Jolie J and Lehmann H 1995 *Nucl Phys* A **593** 21
[3] Délèze M *et al.* 1993 *Nucl Phys* A **554** 1
[4] Garrett P E *et al.* 2007 *Phys Rev* C **75** 054310
[5] Kern J *et al.* 1990 *Nucl Phys* A **512** 1
[6] Wang Y *et al.* 2001 *Phys Rev* C **64** 054315
[7] Aprahamian A *et al.* 1987 *Phys Rev Lett* **59** 535
[8] Wang Y *et al.* 2003 *Phys Rev* C **67** 064303
[9] Garrett P E *et al.* 2008 *Phys Rev* C **78** 044307
[10] Batchelder J C *et al.* 2009 *Phys Rev* C **80** 054318
[11] de Frenne D and Jacobs E 1996 *Nucl Data Sheets* **79** 639 and references therein
[12] Blachot J 2002 *Nucl Data Sheets* **97** 593 and references therein
[13] Brewer N T *et al.* to be published
[14] Batchelder J C *et al.* 2003 *Nucl Instr Meth* B **204** 625
[15] Batchelder J C *et al.* 2005 *Phys Rev* C **72** 044306
[16] Grzywacz R 2003 *Nucl Instr Meth* B **204** 649
[17] Hennig W *et al.* 2007 *Nucl Instr Meth* B **261** 1000
[18] Fogelberg B *et al.* 1971 *Phys Lett* **36**B 334
[19] Jänecke J and Becchetti F D 1979 *Nucl Phys* A **325** 337
[20] Aprahamian A 1985 *Ph D thesis* (Clark University)
[21] Kautzch T *et al.* 2005 *Eur Phys J* A **25** s01 117
[22] Stoyer M A *et al.* 2007 *Nucl Phys* A **787** 455c
[23] Hoteling N *et al.* 2007 *Phys Rev* C **76** 044324
[24] Bargholtz C *et al.* 1975 *Z Phys* A **272** 3

XXXV Symposium on Nuclear Physics IOP Publishing
Journal of Physics: Conference Series **387** (2012) 012006 doi:10.1088/1742-6596/387/1/012006

Superallowed $0^+ \rightarrow 0^+$ beta-decay from $T_z = -1$ sd-shell nuclei

J.C. Hardy and I.S. Towner

Cyclotron Institute, Texas A&M University, College Station, TX 77843-3366

E-mail: hardy@comp.tamu.edu

Abstract. Superallowed nuclear beta-decay between 0^+ analogue states probes the vector part of the weak interaction, with the measured ft-value of each such transition leading to a value for the vector coupling constant, G_V. To date, the ft-values for thirteen $0^+ \rightarrow 0^+$ transitions have been measured with $\sim 0.1\%$ precision or better. The results yield fully consistent values for G_V and an experimental value for V_{ud}, which is the leading diagonal element of the quark mixing matrix, the Cabibbo-Kobayashi-Maskawa (CKM) matrix. With a precise value for V_{ud} established, the unitarity of the CKM matrix can be tested and limits set on the possibility of new physics lying beyond the Standard Model. This work demands high experimental precision and a high degree of reliability in the small theoretical correction terms required to extract G_V and V_{ud} from the experimental data. One of the correction terms must account for isospin symmetry-breaking between the parent and daughter nuclei, and its associated uncertainty contributes significantly to the uncertainty in V_{ud}. The superallowed decays of $T_z = -1$ sd-shell nuclei, such as ^{22}Mg, ^{26}Si, ^{34}Ar and ^{38}Ca, have so far not played a significant role in the determination of V_{ud} because experimental challenges have made high precision unattainable for these transitions. However, if they were to be measured precisely, they would enable important tests of the calculated isospin symmetry-breaking corrections and potentially reduce the latter's uncertainties. Experiments aimed at characterizing these transitions are described.

1. Introduction

Superallowed beta decay between nuclear analogue states with $T = 1$ and $J^\pi = 0^+$ occurs only via the vector current of the weak interaction: angular momentum conservation completely rules out the axial-vector current, which must carry off a spin of one and cannot connect two states that both have spin zero. Furthermore, since the parent and daughter states are analogues of one another, the strength of the transition is affected only by the small difference between the parent and daughter configurations resulting from isospin symmetry breaking, not by the dominant nuclear structure common to them both.

The measured strength of such a transition – expressed as an "ft value" – can then be related directly to the vector coupling constant, G_V with the intervention of only a few small ($\sim$1%) calculated terms to account for radiative and isospin symmetry-breaking effects. Once G_V has been determined in this way, it is only another short step to obtain a value for V_{ud}, the up-down mixing element of the Cabibbo-Kobayashi-Maskawa (CKM) matrix, with which it is possible to test the top-row unitarity of that matrix. Since the unitary CKM matrix is a central pillar of the three-generation Standard Model, any experimentally determined deviation from CKM unitarity would be a signature of new physics beyond the Model; and even uncertainty limits

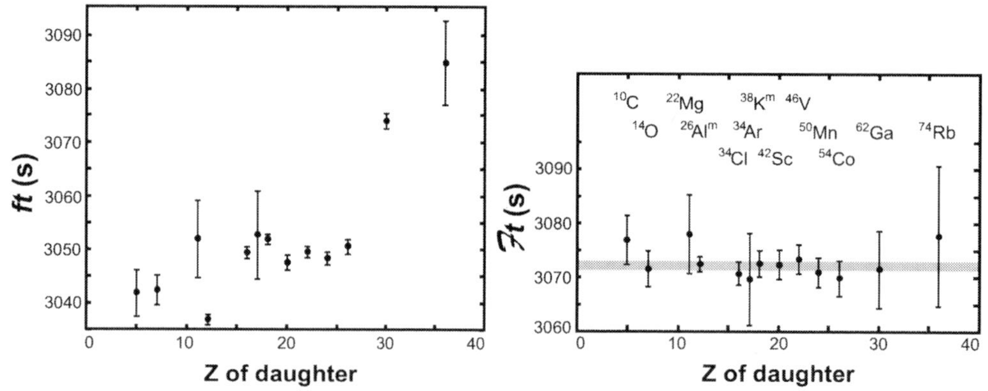

Figure 1. Results from the 2009 survey [1]: uncorrected ft values for the 13 best known superallowed decays on the left; the same results but incorporating the δ'_R, δ_C and δ_{NS} correction terms on the right. The grey band in the right panel is the average $\mathcal{F}t$ value and its uncertainty.

on a sum that agrees with unitarity can serve as a constraint on possible candidates for new physics.

Currently superallowed β-decay yields the most precise value for V_{ud} and the most exacting test of CKM unitarity, with a precision of 0.06% on the latter. This precision can be expected to improve further as a result of decay measurements that focus specifically on defining the effects of isospin-symmetry breaking between the analogue parent and daughter states in each superallowed transition. Previously uncharacterized superallowed transitions from $T_z = -1$ nuclei in the sd-shell can play an important role in this development.

2. Present status of superallowed decays

In dealing with superallowed decays, it is convenient to combine some of the small correction terms with the measured ft-value and define a "corrected" $\mathcal{F}t$-value. Thus, we write [1]

$$\mathcal{F}t \equiv ft(1+\delta'_R)(1+\delta_{NS}-\delta_C) = \frac{K}{2G_{\mathrm{V}}^2(1+\Delta_{\mathrm{R}}^{\mathrm{V}})}\,, \tag{1}$$

where $K = 8120.2787(11) \times 10^{-10}$ GeV^{-4}s; δ_C is the isospin-symmetry-breaking correction and $\Delta_{\mathrm{R}}^{\mathrm{V}}$ is the transition-independent part of the radiative correction. The terms δ'_R and δ_{NS} comprise the transition-dependent part of the radiative correction, the former being a function only of the electron's energy and the Z of the daughter nucleus, while the latter, like δ_C, depends in its evaluation on the details of nuclear structure. From this equation, it can be seen that a measurement of any one superallowed transition establishes an individual value for G_{V}. A measurement of several of them tests the Conserved Vector Current (CVC) hypothesis that G_{V} is not renormalized in the nuclear medium. If indeed G_{V} is constant – *i.e.* all the $\mathcal{F}t$-values are the same – then an average value for G_{V} can be determined and V_{ud} obtained from the relation $V_{ud} = G_{\mathrm{V}}/G_{\mathrm{F}}$, where G_{F} is the well known [2] weak-interaction constant for purely leptonic muon decay.

The ft-value that characterizes any β-transition depends on three measured quantities: the total transition energy, Q_{EC}, the half-life, $t_{1/2}$, of the parent state and the branching ratio, R, for the particular transition of interest. The most recent critical survey of world data on superallowed $0^+ \rightarrow 0^+$ beta decays was published [1] in 2009. It lists the ft values for 13 transitions, which have been precisely determined from a very robust data set with more than 150 independent measurements contributing to the various input quantities. The results were then used to obtain the corrected $\mathcal{F}t$ values, with the outcome shown in Fig. 1.

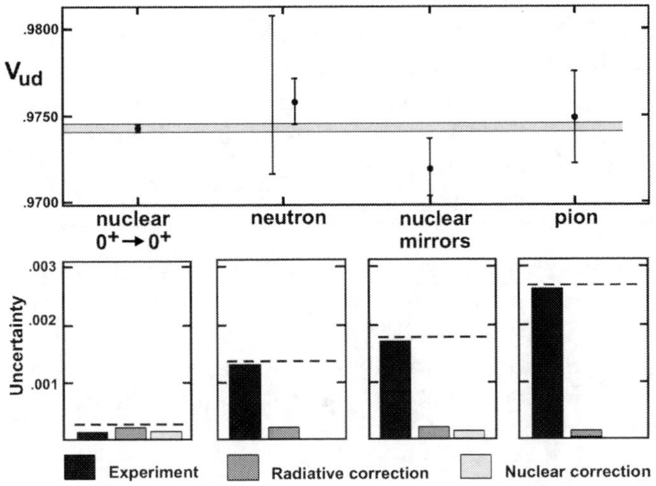

Figure 2. (top) Values for V_{ud} derived from four different types of measurement, the grey band being the average value. The first of the neutron values includes all data; the second includes just the most recent measurements.
(bottom) The four panels show the error budgets for the four results shown in the top panel with points and error bars. The three contributors to the uncertainties – experiment, radiative correction and nuclear correction – are separately identified.

It is immediately evident from the figure that the $\mathcal{F}t$ values are all consistent with one another from A=10 to A=74. This simultaneously confirms the CVC expectation of a constant value for G_V and demonstrates the absence of any significant scalar current, which would introduce an upward or downward curve into the $\mathcal{F}t$-value locus at low Z. It also goes a long way towards validating the calculated isospin symmetry-breaking corrections: The calculations of δ_C and δ_{NS} for each transition in this analysis [3] employed the best available shell-model wave functions, which had been based on a wide range of spectroscopic data for nuclei in the same mass region. They were further tuned to agree with measured binding energies, charge radii and coefficients of the isobaric multiplet mass equation for the specific states involved. This means that the origins of these correction terms are completely independent of the superallowed decay data, so consistency in the corrected $\mathcal{F}t$ values gives powerful support to the calculated corrections used in the derivation of those $\mathcal{F}t$ values.

With a mutually consistent set of $\mathcal{F}t$ values, one is then justified in proceeding to determine the value of G_V and, from it, V_{ud}. The result we obtained in Ref. [1] was V_{ud} = 0.97425(22), which, when combined with current values for V_{us} and V_{ub}, yielded a CKM unitarity sum of 0.99995(61), in remarkable agreement with Standard Model expectations. Note that this analysis was only possible because the $\mathcal{F}t$ values formed a consistent set. Without demonstrated consistency with CVC, there can be no justification for extracting a unique value for G_V from the data, let alone one for V_{ud}.

In the top panel of Fig. 2, the value for V_{ud} obtained from superallowed $0^+ \to 0^+$ β-decays is compared with the values obtained from neutron decay, mirror $T=\frac{1}{2}$ nuclear β decays, and pion decay, all three of which are much less precise but still agree within their quoted uncertainties. The individual contributions to the overall uncertainties for each method are displayed in the bottom panels of the figure. The $0^+ \to 0^+$ result is the only one not dominated by the experimental uncertainty. For it, the radiative correction – principally Δ_R^V – is the largest contributor to the overall uncertainty, with the "nuclear correction" – δ_C and δ_{NS} – a close second.

Although from this analysis of uncertainties one might conclude that experiment has no further role to play in improving the CKM unitarity test via $0^+ \to 0^+$ superallowed decays, this would not be correct. Since the correction terms δ_C and δ_{NS} exhibit very pronounced differences from transition to transition (compare the two panels in Fig. 1, which differ principally by the application of these correction terms) their veracity can be tested and possibly improved by new measurements that either reduce the experimental uncertainties on the currently measured ft

values, or else increase the number of precisely measured transitions, particularly by including cases with much larger calculated correction terms. Depending on whether these new results confirm the transition-to-transition variations obtained from the calculated corrections or not, the calculations may be validated or alternatively refined to restore agreement. Either way, it will likely be possible to reduce the associated theoretical uncertainties.

3. Testing δ_C calculations

We have devised a test [5] that can be applied to any set of isospin symmetry-breaking corrections, δ_C. Our test is based upon the premise that CVC is valid at least to the level of precision attained by the best ft-value measurements. Under that condition, a valid set of structure-dependent correction terms should produce a statistically consistent set of $\mathcal{F}t$ values, the average of which we can write as $\overline{\mathcal{F}t}$. It then follows from Eq. (1) that, for each individual transition in the set, we can write

$$\delta_C - \delta_{NS} = 1 - \frac{\overline{\mathcal{F}t}}{ft(1+\delta'_R)}. \tag{2}$$

For any set of corrections to be acceptable, the calculated value of $\delta_C - \delta_{NS}$ for each superallowed transition must satisfy this equation, where ft is the measured result for that transition and $\overline{\mathcal{F}t}$ has the same value for all of them. Thus, to test a set of correction terms for n superallowed transitions, one can treat $\overline{\mathcal{F}t}$ as a single adjustable parameter and use it to bring the n results from the right side of Eq. (2), which are based predominantly on experiment, into the best possible agreement with the corresponding n calculated values for $\delta_C - \delta_{NS}$. The normalized χ^2, minimized by this process, then provides a figure of merit for that set of calculations.

As it happens, there is only one set of calculations available for δ_{NS} [3, 4] but many for the isospin-symmetry-breaking term δ_C. It therefore becomes more useful to rearrange Eq. (2) to read:

$$\delta_C = 1 + \delta_{NS} - \frac{\overline{\mathcal{F}t}}{ft(1+\delta'_R)}. \tag{3}$$

The same least-squares minimization process can of course be used in the application of this equation.

This test was applied to a number of sets of calculated δ_C correction terms in Ref. [5]. A sample of the results is given here in Fig. 3. Only one theoretical model – the "Shell-model Saxon-Woods" (SM-SW) model [3] illustrated in panel "a" of the figure – produces fully satisfactory agreement with CVC, having a normalized χ^2 of 0.40. A second – the "Shell-model Hartree-Fock" (SM-HF) model [1] shown in panel "b" – is also reasonably acceptable, with a normalized χ^2 of 2.0. It was the first of these, the SM-SW model, that was employed to calculate the δ_C values used in the extraction of V_{ud} from the $0^+ \rightarrow 0^+$ ft-value data [1]. As already described, this is a semi-phenomenological model tuned to match a wide variety of experimental data that is independent of the ft values themselves. The SM-HF model is similarly semi-phenomenological in character, and it was used in a second analysis of the data, in order to assess the extent to which the final result differed between two models that both gave reasonable agreement with CVC. The final uncertainty quoted for V_{ud} in Ref. [1] included a systematic component that reflected the small difference obtained.

A representative of the unacceptable models, the isovector monopole resonance model [6], is presented in panel "c" of the figure; it completely fails to reproduce the data. However, a completely new calculation based on nuclear density functional theory has appeared [7] since our test was published. Its results are given in panel "d". Although its normalized χ^2 is relatively high at 5.2, this is almost entirely due to the disagreement with the ft value for the decay of ^{62}Ga to ^{62}Zn, which is plotted at $Z = 30$. If that transition is removed from consideration, the

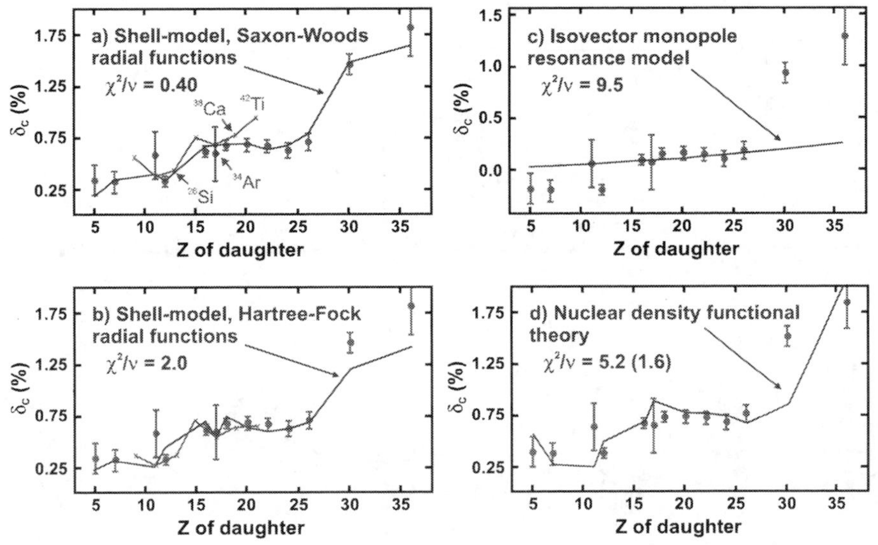

Figure 3. Isospin-symmetry-breaking correction, δ_C, in percent units plotted as a function of atomic number, Z, of the daughter nucleus. The solid circular points with error bars are the values of δ_C obtained from Eq. (3), with the experimental ft values and the values of δ'_R and δ_{NS} (and their uncertainties) all taken from Refs. [1, 5]. In effect, we treat these as the "experimental" δ_C values. The blue lines represent the δ_C values for the well-measured ft values as calculated by the various models described in the text and identified in the upper left of each graph. The value of $\overline{\mathcal{F}t}$ in Eq. (3) has been adjusted in each case by least-squares fitting to optimize the agreement between the "experimental" δ_C values and the calculated ones. The corresponding values of χ^2/n_d are also shown. The green lines in panels "a" and "b" represent the calculated δ_C values for $T_z = -1$ nuclei between ^{18}Ne and ^{42}Ti.

normalized χ^2 becomes 1.6. It is particularly gratifying that the value of V_{ud} extracted from the ft-value data by use of this model [7] agrees well with the result obtained earlier from the SM-SW model [1].

Panels "a" and "b" of Fig. 3 include lines in green that represent the calculated δ_C values for $T_z = -1$ nuclei between ^{18}Ne and ^{42}Ti. Of particular interest are the four parent nuclei specifically identified in panel "a". These are the parents of superallowed transitions from $T_z = -1$ to $T_z = 0$ nuclei, corresponding to each of which there is another superallowed decay from the $T_z = 0$ to $T_z = +1$ nuclei with the same value of A: for example $^{34}\mathrm{Ar} \rightarrow {}^{34}\mathrm{Cl}$ and $^{34}\mathrm{Cl} \rightarrow {}^{34}\mathrm{S}$. By comparing the calculations for these "mirror" transitions in Fig. 3 one can see an interesting difference between the SM-SW calculations and the SM-HF ones. The former consistently predict that the δ_C value for the $T_z = -1$ parent is larger than the δ_C value for the $T_z = 0$ parent, while the SM-HF calculation predicts just the opposite.

In fact, we find for the SM-HF model that the difference between the mirror δ_C values predicted by that model depends systematically on the value of the x_0 asymmetry parameter in the Skyrme potential used in the Hartree-Fock calulation of the radial wave functions. We have calculated the mirror δ_C values using the SM-HF model with 15 different Skyrme potentials taken from the literature, and find a convincing linear dependence between the predicted δ_C difference and the x_0 parameter in the potential [8]. It is even possible to find a potential that predicts very nearly the same result as does the SM-SW model.

As of now, none of the superallowed transitions from the $T_z = -1$ parents identified in Fig. 3 has been measured with sufficient precision to distinguish among these various δ_C predictions.

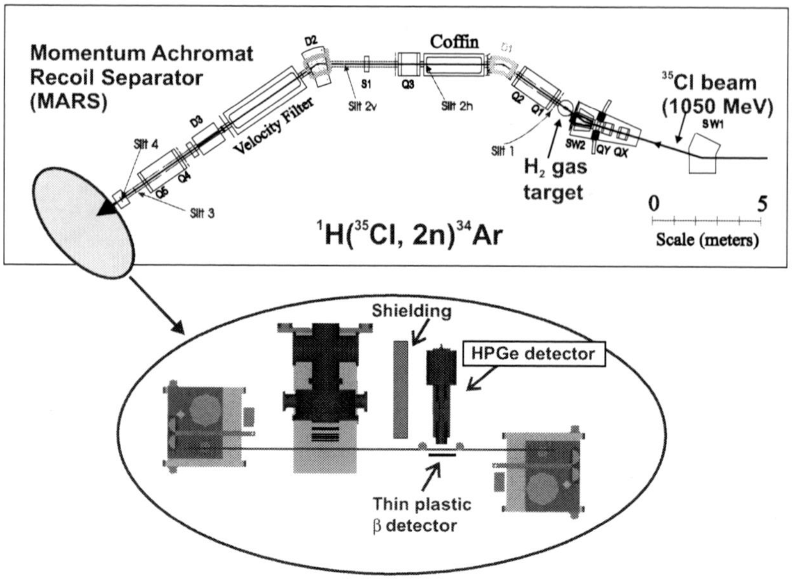

Figure 4. Experimental arrangement used for the ^{34}Ar branching-ratio measurement. For the half-life measurements, the HPGe detector and thin plastic β detector were replaced by a 4π gas proportional counter; the tape passed through the center of that detector.

However, if this deficiency could be rectified it would have several important benefits. It could distinguish definitively between the SM-SW and SM-HF models, thus removing the need to include both when determining the uncertainty on V_{ud}. It might also provide a means for constraining the asymmetry parameter x_0 in the Skyrme potential.

4. Measurements of decays from $T_z = -1$ nuclei

To determine any superallowed transition with useful precision, it is necessary to measure its Q_{EC}-value to ± 100 eV, its half-life to $\pm 0.02\%$ and its branching ratio to $\pm 0.1\%$. For most of the well known $T_z = 0$ parent decays, the branching ratio for the superallowed branch is greater than 99% so it has been determined very precisely via measurement of the small competing Gamow-Teller branches. If the latter can be measured even to $\pm 10\%$, when the result is subtracted from 100% it leads to an uncertainty of less than 0.1% in the superallowed branching ratio.

For $T_z = -1$ parent decays, the situation is much more complicated. First, the branching ratio for the superallowed branch is smaller, and the Gamow-Teller branches are correspondingly larger: for ^{26}Si the superallowed branch is $\sim$76%; for ^{34}Ar it is $\sim$94%; and for ^{38}Ca it is $\sim$77%. Second, each of these decays is immediately followed by the decay of its daughter, which in each case has a half-life about twice that of the decay that feeds it. This means that the total beta-decay curve is composite, with the daughter's half-life predominating. Third, these $T_z = -1$ nuclei are farther from stability than the $T_z = 0$ ones, so are more difficult to produce copiously and free of impurities.

At Texas A&M, we are currently embarked on a program to measure the $T_z = -1$ parent superallowed decays, particularly those like ^{26}Si, ^{34}Ar and ^{38}Ca that feed mirror superallowed decays from their daughters. We have completed and published our half-life results for these three cases [9, 10, 11] and are currently at work on the branching ratios.

Taking ^{34}Ar as an example, we achieve the goal of purity by using a production reaction with inverse kinematics, ^{1}H (^{35}Cl, 2n) ^{34}Ar, and selecting the desired reaction product with the Momentum Achromatic Recoil Separator (MARS). See Fig. 4. A primary beam of 30-AMeV ^{35}Cl from the Texas A&M superconducting cyclotron impinges on a liquid-nitrogen-cooled hydrogen gas target operated at 1.6-atm. The resultant 26-AMeV beam of ^{34}Ar separated by MARS exits the vacuum chamber through a thin Kapton window and then passes through a 0.3-mm-thick plastic scintillator and a series of Al degraders, which are adjusted to ensure the

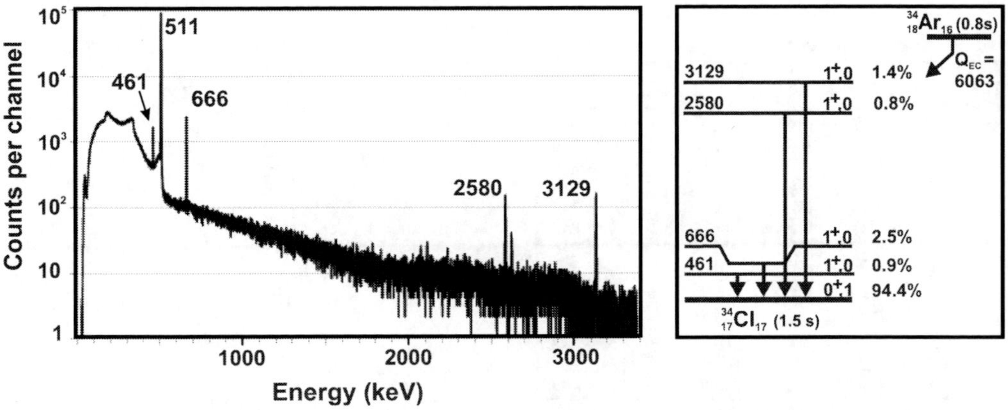

Figure 5. Spectrum of γ rays observed in coincidence with positrons from collected ^{34}Ar sources. Also shown is the β-decay scheme for ^{34}Ar.

implantation of the ^{34}Ar nuclei at the center of a 76-μm-thick aluminized Mylar tape, part of our fast tape-transport system. With an ^{34}Ar beam intensity of about 3×10^4 particles/s, we collect a radioactive sample, >99.8% pure, for typically 1s, then turn off the beam and transport the sample in 180 ms to a shielded counting location where data are collected for a preset length of time, after which the cycle is repeated. This sequence is repeated until sufficient statistics have been collected.

What detection equipment is placed at the counting location depends on whether a branching-ratio or a half-life is being measured. Figure 4 shows the arrangement used for the former, in which a spectrum of γ-rays was recorded for those events in the HPGe detector that were observed to be in coincidence with positrons in the thin plastic β detector. For the latter, a 4π gas proportional counter split into two halves was used, with the tape passing between the halves and the collected sample being stopped exactly at the center for each measurement period [9].

The spectrum of recorded β-delayed γ rays is shown in left panel of Fig. 5, where no impurity peaks can be detected. The β-decay branching ratios were determined from the corresponding intensities of the β-delayed γ-ray peaks, which all correspond to γ-transitions to the ground-state (see the right panel in Fig. 5). If the γ ray de-exciting state i in the daughter is denoted by γ_i, then the β-branching ratio, R_i, for the β-transition populating that state can be written:

$$R_i = \frac{N_{\beta\gamma_i}}{N_\beta \epsilon_{\gamma_i}} k, \tag{4}$$

where $N_{\beta\gamma_i}$ is the total number of β-γ coincidences measured in the γ_i peak, N_β is the total number of β singles, ϵ_{γ_i} is the detector efficiency for γ ray, γ_i, and k is a small correction factor (*i.e.* $k \sim 1$) that, among other things, takes into account the differences in the β-detector efficiency for the different transitions participating in ^{34}Ar decay. (See Ref. [12] for a detailed description of the factors that contribute to k.) This relation highlights the importance of a precise absolute efficiency calibration for the γ-ray detector and a reasonable knowledge of relative efficiencies in the beta detector. Our HPGe detector's absolute efficiency is accurately known (to $\pm0.2\%$ for 50-1400 keV γ rays and $\pm0.4\%$ up to 3500keV) from source measurements and Monte Carlo calculations [13]. The relative efficiency as a function of β energy in the plastic scintillator was determined by Monte Carlo calculations and checked by comparison with measurements on conversion-electron sources.

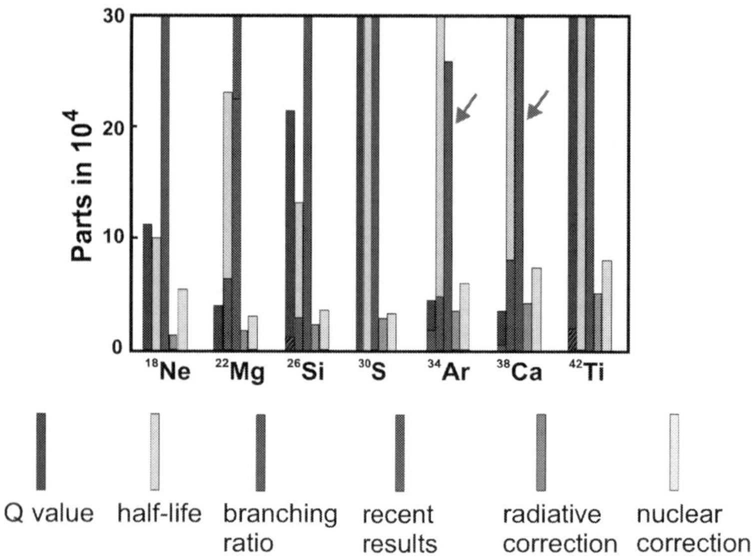

Figure 6. Error budgets for superallowed $0^+ \rightarrow 0^+$ transitions from $T_z = -1$ parent nuclei between ^{18}Ne and ^{42}Ti. The theory contributions are shown in two shades of grey; the three experimental contributions are identified by the color code give at the bottom. Red bars indicate recent improvements in any of the three. The red arrows indicate the two branching ratios we expect to improve significantly in the near future.

Branching-ratio measurements for both ^{34}Ar and ^{38}Ca have been made but are still being evaluated, with special attention directed to dead times and other potential sources of error. We anticipate having final results this year with a precision on the superallowed branch that reaches $\pm 0.2\%$ and perhaps better. Figure 6 presents an overview of the current status of error budgets for the superallowed $0^+ \rightarrow 0^+$ transitions from $T_z = -1$ parent nuclei between ^{18}Ne and ^{42}Ti. The red bars show the improved results that have appeared over the last few years, and the red arrows point to the ^{34}Ar and ^{38}Ca branching ratios we expect to improve soon.

5. Conclusions

Measurements of superallowed $0^+ \rightarrow 0^+$ nuclear beta decays have already achieved remarkable precision and are playing a important role in delimiting the electroweak standard model. Most significantly, they have demonstrated the unitarity of the CKM matrix to within 0.06%. Current activity in the field is now focused on improving this precision even further, with special emphasis on reducing, through experiment, the uncertainties currently attached to the theoretical correction terms that must be applied to the experimental results before fundamental weak interaction parameters can be extracted from them. The isospin-symmetry-breaking corrections are particularly important, but we can test the validity of any set of such calculations by requiring that the resulting corrected $\mathcal{F}t$ values satisfy the conservation of the vector current by being statistically consistent with one another from transition to transition.

We have described here a potentially powerful way to apply this test to mirror pairs of $0^+ \rightarrow 0^+$ transitions: for example ^{34}Ar $\rightarrow$ ^{34}Cl and ^{34}Cl $\rightarrow$ ^{34}S. All experimental data required for a precise test with this pair of transitions is in hand with the exception of the branching ratio for the decay from ^{34}Ar. An experiment has been described, which is designed to precisely measure this branching ratio, and those like it from other $T_z = -1$ superallowed parent nuclei. Results can be expected within the next year.

We are optimistic that the value of V_{ud} and the CKM unitarity test can still be improved by nuclear measurements.

5.1. Acknowledgments

The experimental work referred to here is being done in collaboration with V.E. Iacob, H.I Park and the other members of the Cyclotron Institute weak-interactions group. The work was supported by the U.S. Department of Energy under contract no. DE-FG02-93ER-40773 and by the Robert A. Welch Foundation under grant no. A-1397.

References

[1] Hardy J C and Towner I S 2009 Phys. Rev. C **79** 055502.
[2] Nakamura K *et. al.* 2010 [Particle Data Group] J. Phys. G **37** 075021 and 2011 partial update for the 2012 edition.
[3] Towner I S and Hardy J C 2008 Phys. Rev. C **77** 025501.
[4] Towner I S 1994 Phys. Lett. B **333** 13.
[5] Towner I S and Hardy J C 2010 Phys. Rev. C **82** 065501.
[6] Auerbach N 2009 Phys. Rev. C **79** 035502.
[7] Satula W, Dobaczewski J, Nazarewicz W and Rafalski M 2011 Phys. Rev. Lett. **106** 132502.
[8] Towner I S and Hardy J C 2012 To be published.
[9] Iacob V E, Hardy J C, Brinkley J F, Gagliardi C A, Mayes V E, Nica N, Sanchez-Vega M, Tabacaru, G, Trache L and Tribble R E 2006 Phys. Rev. C **74** 055502.
[10] Iacob V E, Hardy J C, Banu A, Chen L, Golovko V V, Goodwin J, Horvat V, Nica N, Park H I, Trache L and Tribble R E 2010 Phys. Rev. C **82** 035502.
[11] Park H I, Hardy J C, Iacob V E, Banu A, Chen L, Golovko V V, Goodwin J, Horvat V, Nica N, Simmons E, Trache L and Tribble R E 2011 Phys. Rev. C **84** 065502.
[12] Iacob V E, Hardy J C, Gagliardi C A, Goodwin J, Nica N, Park H I, Tabacaru G, Trache L, Tribble R E, Zhai Y and Towner I S, Phys. Rev. C **74**, 015501 (2006).
[13] Hardy J C, Iacob V E, Sanchez-Vega M, Effinger R T, Lipnik P, Mayes V E, Willis D K and Helmer R G, Appl. Radiat. Isot. **56**, 65 (2002); Helmer R G, Hardy J C, Iacob V E, Sanchez-Vega M, Neilson R G and Nelson J, Nucl. Instr. Meth. Phys. Res. A **511**, 360 (2003); and Helmer R G, Nica N, Hardy J C and Iacob V E, Appl. Radiat. Isot. **60**, 173 (2004).

Journal of Physics: Conference Series **387** (2012) 012007 doi:10.1088/1742-6596/387/1/012007

Nuclear and weak interaction aspects of neutrinoless double beta decay: recent results

O Civitarese
Department of Physics. University of La Plata, c.c.67 (1900) La Plata, Argentina
E-mail: osvaldo.civitarese@fisica.unlp.edu.ar

J Suhonen
Department of Physics. University of Jyväskylä, P.O.Box 35 (YFL) Fl-40014, Finland
E-mail: jouni.suhonen@phys.jyu.fi

Abstract. The determination of the value of the light neutrino mass, as well as the determination of the nature of the neutrino, are two of the fundamental questions which motivate the experimental search of signals of neutrinoless double beta decay transitions. Here, we shall review some of the the essentials of the theory, based on both nuclear structure and elementary particle physics, relevant for the understanding of the problem

1. Introduction

Various experiments conducted during the last decade have confirmed the existence of neutrino oscillations [1]. The experimental findings have therefore confirmed the theoretical notions advanced by Pontecorvo many years ago. The nature of the neutrino (as a Majorana or Dirac particle) was not established yet. It requires to pass a more challenging test, that is the experimental observation of neutrinoless double beta decay [2] and the determination of an associated lepton-number violating mechanism [3].

The double beta decay is a very rare decay which takes place between nuclei which differ in two units of charge and that have the same mass number; $(A, N, Z) \rightarrow (A, N \mp 2, Z \pm 2)$; and it produces two electrons and two antineutrinos (two-neutrino mode) or just two-electrons (neutrinoless mode). The two-neutrino mode is allowed by the Standard Model of Electroweak interactions since it conserves the lepton number, but it is suppressed by kinematical reasons (e.g.: four leptons in the final state) and it is independent of the neutrino properties. The neutrinoless mode is by far the most interesting one, because it is forbidden by the Standard Model of Electroweak interactions (e.g.; it implies lepton number violation), but if it is detected, it will demonstrate that lepton number is not a fundamental symmetry. The two neutrino mode has been observed in direct measurements, for a series of nuclei and it has the longest half-life ($\approx 10^{20}$ years) ever detected directly (that is in a laboratory). The expected half-life for the neutrinoless mode is even larger, of the order of $> 10^{25}$ years [2].

The calculation of nuclear matrix elements (NME) for double beta decay (DBD) transitions is a matter of relevance, both for nuclear and particle physics. We shall referred hereafter to [4, 5] for details. It seems that a consensus has been reached concerning the order of magnitude

Journal of Physics: Conference Series **387** (2012) 012007 doi:10.1088/1742-6596/387/1/012007

of the relevant NME [5], and that the reliability of the theoretical estimations of NME has improved. In this note we review some of the elements which, in our opinion, are crucial to determine the stability of the results. For details about the calculations we shall refer the reader to Refs.[6, 7, 8].

2. Neutrino mixing

The relation between flavor ($\nu_{e,\mu,\tau}$) and mass ($\nu_{1,2,3}$) eigenstates, of the neutrino, assuming light neutrinos and CP invariance is written

$$\begin{pmatrix} \nu_e \\ \nu_\mu \\ \nu_\tau \end{pmatrix} = U \otimes \begin{pmatrix} \nu_1 \\ \nu_2 \\ \nu_3 \end{pmatrix} \tag{1}$$

The elements of the mixing matrix U can be determined from oscillation data, and the mass eigenstates can be labelled by hierarchies of the mass values (either by adopting a normal ($m_1 \approx m_2 < m_3$), inverse ($m_3 < m_1 \approx m_2$) or degenerate ($m_1 \approx m_2 \approx m_3$) values). The square mass differences $\delta^2_{ij} = m^2_i - m^2_j$ are fixed by solar and atmospheric neutrino-oscillation data data [1], and the value of the sum of the masses, $\Omega = \sum_i m_i$, may be extracted from astronomical data [9]. The absolute mass scale can be fixed by DBD, once the decay mechanism is determined by data. The average electron-neutrino mass is written

$$< m_\nu >= \sum_{i=1}^{3} m_i U^2_{ei} \lambda^{CP}_i = m_1 U^2_{e1} \pm m_2 U^2_{e2} \tag{2}$$

(we assume CP invariance and $U_{13} = 0$). The average electron-neutrino mass (in units of eV), for the best fit-matrix U and for the three mass hierarchies[6] may vary between 0.10 eV and 0.50 eV.

3. Nuclear structure

The mass mode of the $0\nu\beta\beta$ decay describes the exchange of a light virtual Majorana neutrino between two decaying neutrons of the initial nucleus. The $0\nu\beta\beta$-decay half-life is related to the effective neutrino mass $\langle m_\nu \rangle$ through the relation

$$\left[t^{(0\nu)}_{1/2}(0^+_{\rm i} \rightarrow 0^+_{\rm f}) \right]^{-1} = G^{(0\nu)}_1 \left(\frac{\langle m_\nu \rangle}{m_{\rm e}} \right)^2 \left(M^{(0\nu)} \right)^2 , \tag{3}$$

where $G^{(0\nu)}_1$ is the leptonic phase-space factor [3]. The $0\nu\beta\beta$ nuclear matrix element $M^{(0\nu)}$ consists of the Gamow–Teller (GT), Fermi (F) and tensor (T) parts as

$$M^{(0\nu)} = M^{(0\nu)}_{\rm GT} - \left(\frac{g_{\rm V}}{g_{\rm A}} \right)^2 M^{(0\nu)}_{\rm F} + M^{(0\nu)}_{\rm T} . \tag{4}$$

The tensor part in (4) is very small, and it can be neglected. The explicit expression of these matrix elements can be found in [6]. The review of the fundamentals of the currently used models can be found in [4, 5] and for the sake of brevity we shall focuss here on the spherical proton-neutron Quasiparticle Random Phase Approximation (pnQRPA) and on the discussion of the theoretical assumptions may affect the results.

Journal of Physics: Conference Series **387** (2012) 012007 doi:10.1088/1742-6596/387/1/012007

3.1. The pnQRPA approximation

The pnQRPA consists of the diagonalization of the residual proton-neutron two body interaction in a subspace of two-quasiparticle (proton-neutron) states. It contains a free parameter, the particle-particle strength parameter, $g_{\rm pp}$, that controls the magnitude of the proton-neutron two-body interactions in the $J^\pi = 1^+$ channel [10]. The value of this parameter can be fixed by fitting the data on the available two-neutrino double beta ($2\nu\beta\beta$) decay [6] or data on single beta decay [11]. However, since the correlations which are involved in the NME for two-neutrino ($2\nu\beta\beta$) and neutrinoless ($0\nu\beta\beta$) decays are basically different (see next section), the adjustment of this parameter to the observed $2\nu\beta\beta$ may not guarantee the accuracy of the NME for the neutrinoless DBD mode. The non-physical effects associated to attempts to go beyond the point of collapse of the proton-neutron Quasiparticle Random Phase approximation (pnQRPA) are notorious. As a consequence, the results of some extensions , like the pnRQRPA, should be taken with care.

3.2. Jastrow and UCOM-Short Range Correlations

The average exchanged momentum, carried by the neutrino is of the order of 100 MeV/c, and the two nucleons tend to overlap. This is a very distinctive feature of the NME for the neutrinoless double beta decay (the $0\nu\beta\beta$ NME are relatively large, of the order of 3-5) as compared with the two-neutrino mode (the $2\nu\beta\beta$ NME are very small, e.g: ten to hundred times smaller than the $0\nu\beta\beta$ NME). A suitable microscopic approach for the inclusion of short-range correlations is the unitary correlation operator method (UCOM). In [7] it was demonstrated that the conventional Jastrow procedure leads to excessive reductions in the magnitudes of the $0\nu\beta\beta$ nuclear matrix elements. For the Jastrow function, the combined action of the exponential and polynomial factors (which are functions of the relative distance between nucleons) affects the balance between the negative and positive contributions of the radial part of the two-nucleon wave function, suppressing the predominantly positive contribution to the nuclear matrix element. The UCOM correlation function acts upon the surface portion of the two-particle wave function, thus producing a slighter reduction. A consistent description of short-range correlations should include, at the same level of approximation, the treatment of the wave functions and that of the transition operators. This can be achieved by constructing a theory of effective operators. The differences which emerge from the study of the radial dependence of the NME is another element to be considered at the time of adopting the adjustment of the parameter g_{pp}.

Here, as illustration of these concepts, we present the results of nuclear-structure calculations for the $0\nu\beta\beta$ ground-state-to-ground-state decays of ^{76}Ge, ^{82}Se, ^{96}Zr, ^{100}Mo, ^{116}Cd, ^{128}Te, ^{130}Te, and ^{136}Xe [7]. To obtain physical values of the proton-neutron particle-particle interaction strength, $g_{\rm pp}$, of the pnQRPA, it was adjusted to reproduce the experimental rates of the $2\nu\beta\beta$ decay. The fit included the experimental errors and the uncertainty in the value of the axial-vector coupling constant $g_{\rm A}$ ($1.0 \le g_{\rm A} \le 1.25$). The extracted values of $g_{\rm pp}$ and the corresponding values of $g_{\rm A}$ are listed in Table 1(see [7] for further details).

4. Adjusted single-particle energies and BCS occupation factors

The results of charge-exchange and one particle transfer measurements in A=76 nuclei can be used as a guideline to adjust single-particle energies and occupation factors [12]. The proton energies were inspected by using the odd-mass nuclei adjacent to ^{76}Ge and ^{76}Se as spectroscopic tools [8]. The standard method to determine single-particle energies is to use the Woods–Saxon mean-field potential in conjunction with spectroscopic data. Small adjustments of the resulting energies can be done based on the data, particularly by looking at the sequence of energy levels of odd-mass nuclei in the neighborhood of the nucleus where the pnQRPA calculation are

Table 1. Calculated $0\nu\beta\beta$ NME for some of the DBD emitters. The $g_{\rm pp}$ and $g_{\rm A}$ values and the resulting half-lives $t^{(0\nu)}_{1/2}$, are given in the table. Jastrow (J) and UCOM (U) matrix elements are given in the last columns. The half-lives $t^{(0\nu)}_{1/2}$ are expressed in units of yr/($\langle m_\nu \rangle$[eV])^2.

Nucleus	$g_{\rm pp}$	$g_{\rm A}$	$M^{(0\nu)}$(J)	$t^{(0\nu)}_{1/2}$(J)	$M^{(0\nu)}$(U)	$t^{(0\nu)}_{1/2}$(U)
^{76}Ge	1.02	1.00	5.077	4.0×10^{24}	6.555	2.4×10^{24}
	1.06	1.25	4.029	2.6×10^{24}	5.355	1.4×10^{24}
^{82}Se	0.96	1.00	3.535	1.9×10^{24}	4.597	1.1×10^{24}
	1.00	1.25	2.771	1.2×10^{24}	3.722	6.9×10^{23}
^{96}Zr	1.06	1.00	3.131	1.2×10^{24}	4.319	6.1×10^{23}
	1.11	1.25	2.065	1.1×10^{24}	3.117	4.7×10^{23}
^{100}Mo	1.07	1.00	3.526	1.2×10^{24}	4.849	6.2×10^{23}
	1.09	1.25	2.737	7.9×10^{23}	3.931	3.8×10^{23}
^{116}Cd	0.82 (β^- decay)	1.25	3.981	3.5×10^{23}	4.928	2.3×10^{23}
	0.97	1.00	3.681	1.0×10^{24}	4.682	6.3×10^{23}
	1.01	1.25	3.034	6.1×10^{23}	3.935	3.6×10^{23}
^{128}Te	0.86 (β^- decay)	1.25	4.068	9.5×10^{24}	5.509	5.2×10^{24}
	0.89	1.00	4.279	2.1×10^{25}	5.841	1.1×10^{25}
	0.92	1.25	3.383	1.4×10^{25}	4.790	6.9×10^{24}
^{130}Te	0.84	1.00	4.061	9.5×10^{23}	5.442	5.3×10^{23}
	0.90	1.25	2.993	7.0×10^{23}	4.221	3.5×10^{23}
^{136}Xe	0.74	1.00	2.864	1.8×10^{24}	3.719	1.1×10^{24}
	0.83	1.25	2.053	1.4×10^{24}	2.802	7.6×10^{23}

performed. In [12] the neutron pair correlations in the ^{76}Ge and ^{76}Se nuclei were measured by (p,t) reactions. As a model space we have used the $N = 3$ and $N = 4$ oscillator shells and the $0h_{11/2}$ single-particle orbital, both for protons and neutrons. The single-particle energies were obtained from the Coulomb-corrected Woods–Saxon potential. In Ref. [6, 7, 8] we describe the details of the systematics.

4.1. DBD-NME results

The calculated NME ($0\nu\beta\beta$ mode), for the decay of some DBD systems are shown in Table 2.

Table 2. NME for neutrinoless DBD transitions from the ground state of the mother nuclei. The NME are given for two extreme values of the axial-vector coupling g_A. The factor $C^{(0\nu)}$ contains the phase space factor and the calculated NME [8].

System	NME($g_A = 1$)	NME ($g_A = 1.25$)	$C^{(0\nu)}$
^{76}Ge	3.23	5.52	1.36-3.96
^{82}Se	2.77	4.57	0.46-1.24
^{128}Te	3.74	5.62	4.98-11.2
^{130}Te	3.48	5.12	0.24-0.52
^{136}Xe	2.38	3.35	0.53-1.06

The magnitude of the pnQRPA (UCOM) calculated NME is rather close to the shell-model result. The reason for the reduction of the magnitude of the $0\nu\beta\beta$ NME, as compared with previous pnQRPA results, can be found by performing the multipole decomposition of the NME [8]. For the Fermi matrix element the reduction stems from the 0^+ intermediate states. For the Gamow–Teller matrix element $M^{(0\nu)}_{\rm GT}$ the significant changes concentrate on the 1^+ and 2^- contributions. The wave function of the 2^-_1 state plays a key role when seeking the reason for the reduction of the magnitude of the Gamow–Teller matrix element.

Concerning the decay of ^{76}Ge, the quality of the lowest 2^- state in the intermediate nucleus ^{76}As can be tested by computing the β^- decay $\log ft$ values for transitions from this state to the ground state and one- and two-phonon states in ^{76}Se. This transition tests exclusively the 2^-_1 wave function whereas the rest of the transitions depend also on wave function of the final state. The single β^- decay is a non-trivial way to check the reduction of the $0\nu\beta\beta$ NME. In the calculation of [8] the wave function of the 2^-_1 state is more fragmented and thus reduces the pnQRPA amplitude responsible for the transitions.

The computed $0\nu\beta\beta$ NME of Table 2 can be converted to half-life limits,

$$t^{(0\nu)}_{1/2} \quad = \quad C^{(0\nu)} \times 10^{24}\,{\rm yr}/(\langle m_\nu\rangle[{\rm eV}])^2 \tag{5}$$

which yield, for an average electron-neutrino mass of the order of 0.1 eV, half-life lower limits of the order of 10^{26} years. These limits are of the order of magnitude of the values which may be accessible to the next generation of DBD experiments.

5. Conclusions

To conclude, we have presented the results of pnQRPA-based calculations of the nuclear matrix elements involved in the neutrinoless double beta decay, and discuss some of the aspects related to the sensitivity of the NME. Particularly, we have discussed the use of the available data on particle transfer to adjust the single particle sector of the calculations, and the sensitivity of the results upon the method adopted to account for short range correlations.

Acknowledgments

This work has been partially supported by the CONICET of Argentina (PIP 0740) and by ANPCYT of Argentina, and by the Finnish Center of Excellence Program 2006-2011 (Nuclear and Accelerator Base Program at the JYFL, Finland).

References

[1] SNO Collaboration, Ahmad Q R *et al.* 2002 *Phys Rev Lett* **89** 011301, SK-Collaboration, Fukuda S *et al.* 2001 *Phys Rev Lett* **86** 5651, KamLAND Collaboration, Eguchi K *et al.* 2003 *Phys Rev Lett* **90** 021802, Appollonio M *et al.* 1999 *Phys Lett B* **466** 415
[2] Avignone F T, Elliott S R, Engel J 2008 *Rev Mod Phys* **80** 481
[3] Vergados J D 2002 *Phys Rep* **361** 1, Doi M, Kotani T, Takasugi E 1985 *Prog Theor Phys Suppl* **83** 1
[4] Suhonen J and Civitarese O 1998 *Phys Rep* **300** 123; Faessler A and Simkovic F 1998 *J Phys G* **24** 2139
[5] Barea J and Iachello F 2009 *Phys Rev C* **79** 044301; Suhonen J and Kortelainen M 2008 *Int J Mod Phys E* **17** 1; Simkovic F, Faessler A and Vogel P 2009 *Phys Rev C* **79** 015502; Simkovic F, Faessler A, Rodin V, Vogel P, Engel J 2008 *Phys Rev C* **77** 045503
[6] Civitarese O and Suhonen J 2005 *Phys Lett B* **626** 80; Civitarese O and Suhonen J 2005 *Nucl Phys A* **761** 313; Civitarese O and Suhonen J 2003 *Nucl Phys A* **729** 867
[7] Kortelainen M, Civitarese O, Suhonen J, Toivanen J 2007 *Phys Lett B* **647** 128; Kortelainen M, Suhonen J 2007 *Phys Rev C* **75** 051303(R); Kortelainen M, Suhonen J 2007 *Phys Rev C* **76** 024315
[8] Suhonen J and Civitarese O 2010 *Nucl Phys A* **847** 207
[9] Larson D *et al.* 2011 *Ap J S* **192** 16
[10] Vogel P and Zirnbauer M 1986 *Phys Rev Lett* **57** 3148; Civitarese O, Faessler A and Tomoda T 1987 *Phys Lett B* **194** 11
[11] Suhonen J 1993 *Nucl Phys A* **563** 205
[12] Schiffer J P *et al.* 2008 *Phys Rev Lett* **100** 112501; Kay B P *et al.* 2009 *Phys Rev C* **79** 021301(R)

Journal of Physics: Conference Series **387** (2012) 012008 doi:10.1088/1742-6596/387/1/012008

The neutrino mixing angle θ_{13} in an S_3 flavour symmetric model

F. González Canales[1,2,*] and A. Mondragón[1,]**

[1]Instituto de Física, Universidad Nacional Autónoma de México, 04510, México D.F., México
[2]Facultad de Ciencias de la Electrónica, Benemérita Universidad Autónoma de Puebla, Apdo. Postal 157, 72570, Puebla, Pue., México

E-mail: *ffelix@fisica.unam.mx, **mondra@fisica.unam.mx

Abstract. In this work, we discuss neutrino masses and mixings in the framework of a minimal S_3 symmetric extension of the Standard Model. In this model, the mass matrices of all fermions take the same generic form with two texture zeroes. The mass matrices of the neutrinos and charged leptons are re-parameterized in terms of their eigenvalues, then the neutrino mixing matrix, V_{PMNS}, is computed and exact, explicit analytical expressions for the neutrino mixing angles as functions of the masses of neutrinos and charged leptons are obtained in excellent agreement with the latest experimental data.

1. Introduction

The observation of flavour oscillations of solar, atmospheric, reactor, and accelerator neutrinos established that they have non-vanishing masses and mix among themselves, much like the quarks do [1]. In these observations and experiments, the differences of the squared neutrino masses, as well as the neutrino mixing angles are measured. These discoveries brought out very forcibly the need of extending the Standard Model (SM) in order to accomodate in the theory the new data on neutrino physics in a consistent way that would allow for a unified and systematic treatment of the observed hierarchy of masses and mixings of all fermions. At the same time, the number of free parameters in the extended form of the SM had to be drastically reduced in order to give predictive power to the theory. These two seemingly contradictory demands are met by a flavour symmetry under which the families transfom in a non-trivial fashion. The observed pattern of neutrino mixing and, in particular, the non vanishing and sizable value of the reactor mixing angle strongly suggests a flavour permutational symmetry S_3.

The results of a combined analysis of all available neutrino oscillation data, including the recent results from long-baseline $\nu_\mu \longrightarrow \nu_e$ searches at the Tokai to Kamioka (T2K) [2] and Double CHOOZ experiments [3], as well as, the Main Injector Neutrino Oscillation Search (MINOS) experiment, give the following values for the differences of the squared neutrino masses and the mixing angles in the lepton mixing matrix, U_{PMNS}, at 1σ confidence level [4]:

$$\Delta m_{21}^2 = 7.59^{+0.20}_{-0.18} \times 10^{-5}\ \mathrm{eV}^2, \quad \Delta m_{31}^2 = \begin{cases} -2.40^{+0.08}_{-0.09} \times 10^{-3}\ \mathrm{eV}^2. \\ \\ +2.50^{+0.09}_{-0.16} \times 10^{-3}\ \mathrm{eV}^2. \end{cases} \qquad (1)$$

Journal of Physics: Conference Series **387** (2012) 012008 doi:10.1088/1742-6596/387/1/012008

$$\sin^2\theta^l_{12} = 0.312^{+0.017}_{-0.015}, \quad \sin^2\theta^l_{23} = \begin{cases} 0.52 \pm 0.06, \\ 0.52^{+0.06}_{-0.07}, \end{cases}, \quad \sin^2\theta^l_{13} = \begin{cases} 0.016^{+0.008}_{-0.006}, \\ 0.013^{+0.007}_{-0.005}, \end{cases}, \tag{2}$$

the upper (lower) row corresponds to inverted (normal) neutrino mass hierarchy, see also Gonzalez Garcia *et al* [5] and J. F. W. Valle *et al* [6]. Recently, the Daya Bay [7] and RENO experiments [8] found the following values for the reactor neutrino mixing angle: $\sin^2 2\theta^l_{13} = 0.092 \pm 0.016\,(\text{stat}) \pm 0.005\,(\text{syst})$ which is equivalent to $\theta^l_{13} \simeq 8.8^\circ \pm 0.8^\circ$ at $5.2\,\sigma$ level, and $\sin^2 2\theta^l_{13} = 0.113 \pm 0.013\,(\text{stat}) \pm 0.019\,(\text{syst})$ which is equivalent to $\theta^l_{13} \simeq 9.8^\circ$ at $4.9\,\sigma$, respectively.

In the last ten years, important theoretical advances have been made in the understanding of the mechanism for fermion mass generation and flavour mixing [9, 10, 11]. A phenomenologically and theoretically meaningful approach for reducing the number of free parameters in the Standard Model and its extensions is the imposition of flavour symmetries and/or texture zeroes [9, 10]. In this approach, certain texture zeroes may be obtained from a flavour symmetry [12, 13]. For recent review of flavour symmetry models see [11]. In the case of the Minimal S_3-Invariant Extension of the Standard Model [14, 15, 16, 17, 18, 19, 20], the concept of flavour and generations is extended to the Higgs sector in such a way that all the matter fields – Higgs, quarks, and lepton fields, including the right-handed neutrino fields– have three species and transform under the flavour symmetry group as the three dimensional representation $1 \oplus 2$ of the permutational group S_3. A model with more than one Higgs $SU(2)$ doublet has tree level flavour changing neutral currents whose exchange may give rise to lepton flavour violating processes and may also contribute to the anomalous magnetic moment of the muon [14, 17]. An effective test of the phenomenological success of the model is obtained by verifying that all flavour changing neutral current processes and the magnetic anomaly of the muon, computed in the S_3–Invariant extended form of the Standard Model, agree with the experimental values [17, 18].

2. The Minimal S_3-invariant Extension of the Standard Model

In the Standard Model analogous fermions in different generations have identical couplings to all gauge bosons of the strong, weak and electromagnetic interactions. Prior to the introduction of the Higgs boson and mass terms, the Lagrangian is chiral and invariant with respect to permutations of the left and right fermionic fields, that is, it is invariant under the action of the group of permutations acting on the flavour indices of the matter fields [14, 17].

The six possible permutations of three objects (f_1, f_2, f_3) are elements of the permutational group S_3. This is the discrete, non-Abelian group with the smallest number of elements. The three-dimensional real representation is not an irreducible representation of S_3. It can be decomposed into the direct sum of a doublet f_D and a singlet f_s, where

$$f_s = \tfrac{1}{\sqrt{3}}(f_1 + f_2 + f_3), \quad f_D^T = \left(\tfrac{1}{\sqrt{2}}(f_1 - f_2), \tfrac{1}{\sqrt{6}}(f_1 + f_2 - 2f_3) \right). \tag{3}$$

The direct product of two doublets $\mathbf{p_D}^T = (p_{D1}, p_{D2})$ and $\mathbf{q_D}^T = (q_{D1}, q_{D2})$ may be decomposed into the direct sum of two singlets $\mathbf{r_s}$ and $\mathbf{r_{s'}}$, and one doublet $\mathbf{r_D}^T$ where

$$\mathbf{r_s} = p_{D1}q_{D1} + p_{D2}q_{D2}, \quad \mathbf{r_{s'}} = p_{D1}q_{D2} - p_{D2}q_{D1}, \tag{4}$$

$$\mathbf{r_D}^T = (r_{D1}, r_{D2}) = (p_{D1}q_{D2} + p_{D2}q_{D1}, p_{D1}q_{D1} - p_{D2}q_{D2}). \tag{5}$$

The antisymmetric singlet $\mathbf{r_{s'}}$ is not invariant under S_3 [14, 17].

Since the Standard Model has only one Higgs $SU(2)_L$ doublet, which can only be an S_3 singlet, it can only give mass to the quark or charged leptons in the S_3 singlet representation, one in each family, without breaking the S_3 symmetry [13, 21].

Hence, in order to impose S_3 as a fundamental symmetry, unbroken at the Fermi scale, we are led to extend the Higgs sector of the theory. The quark, lepton and Higgs fields are

$$Q^T = (u_L, d_L) \ , \ u_R \ , \ d_R \ , \ L^T = (\nu_L, e_L) \ , \ e_R \ , \ \nu_R \ \text{and} \ H, \tag{6}$$

in an obvious notation. All of these fields have three species, and we assume that each one forms a reducible representation $\mathbf{1}_S \oplus \mathbf{2}$. The doublets carry capital indices I and J, which run from 1 to 2, and the singlets are denoted by

$$Q_3, \ u_{3R}, \ d_{3R}, \ L_3, \ e_{3R}, \ \nu_{3R}, \ \text{and}\, H_S. \tag{7}$$

Note that the subscript 3 denotes the singlet representation and not the third generation. The most general renormalizable Yukawa interactions of this model are given by [14, 17]

$$\mathcal{L}_Y = \mathcal{L}_{Y_D} + \mathcal{L}_{Y_U} + \mathcal{L}_{Y_E} + \mathcal{L}_{Y_\nu}, \tag{8}$$

where

$$\begin{aligned} \mathcal{L}_{Y_D} \ = \ & -Y_1^d \overline{Q}_I H_S d_{IR} - Y_3^d \overline{Q}_3 H_S d_{3R} - Y_2^d [\ \overline{Q}_I \kappa_{IJ} H_1 d_{JR} + \overline{Q}_I \eta_{IJ} H_2 d_{JR}\ ] \\ & -Y_4^d \overline{Q}_3 H_I d_{IR} - Y_5^d \overline{Q}_I H_I d_{3R} + \text{ h.c.}, \end{aligned} \tag{9}$$

$$\begin{aligned} \mathcal{L}_{Y_U} \ = \ & -Y_1^u \overline{Q}_I (i\sigma_2) H_S^* u_{IR} - Y_3^u \overline{Q}_3 (i\sigma_2) H_S^* u_{3R} - Y_4^u \overline{Q}_3 (i\sigma_2) H_I^* u_{IR} \\ & -Y_2^u [\ \overline{Q}_I \kappa_{IJ} (i\sigma_2) H_1^* u_{JR} + \eta \overline{Q}_I \eta_{IJ} (i\sigma_2) H_2^* u_{JR}\ ] - Y_5^u \overline{Q}_I (i\sigma_2) H_I^* u_{3R} + \text{ h.c.}, \end{aligned} \tag{10}$$

$$\begin{aligned} \mathcal{L}_{Y_E} \ = \ & -Y_1^e \overline{L}_I H_S e_{IR} - Y_3^e \overline{L}_3 H_S e_{3R} - Y_2^e [\ \overline{L}_I \kappa_{IJ} H_1 e_{JR} + \overline{L}_I \eta_{IJ} H_2 e_{JR}\ ] \\ & -Y_4^e \overline{L}_3 H_I e_{IR} - Y_5^e \overline{L}_I H_I e_{3R} + \text{ h.c.}, \end{aligned} \tag{11}$$

$$\begin{aligned} \mathcal{L}_{Y_\nu} \ = \ & -Y_1^\nu \overline{L}_I (i\sigma_2) H_S^* \nu_{IR} - Y_3^\nu \overline{L}_3 (i\sigma_2) H_S^* \nu_{3R} - Y_4^\nu \overline{L}_3 (i\sigma_2) H_I^* \nu_{IR} \\ & -Y_2^\nu [\ \overline{L}_I \kappa_{IJ} (i\sigma_2) H_1^* \nu_{JR} + \overline{L}_I \eta_{IJ} (i\sigma_2) H_2^* \nu_{JR}\ ] - Y_5^\nu \overline{L}_I (i\sigma_2) H_I^* \nu_{3R} + \text{ h.c.}, \end{aligned} \tag{12}$$

and

$$\kappa = \begin{pmatrix} 0 & 1 \\ 1 & 0 \end{pmatrix} \quad \text{and} \quad \eta = \begin{pmatrix} 1 & 0 \\ 0 & -1 \end{pmatrix}. \tag{13}$$

Furthermore, we add to the Lagrangian the Majorana mass terms for the right-handed neutrinos

$$\mathcal{L}_M = -M_1 \nu_{1R}^T C \nu_{1R} - M_2 \nu_{2R}^T C \nu_{2R} - M_3 \nu_{3R}^T C \nu_{3R}, \tag{14}$$

where M_i with $i = 1, 2, 3$ are the right-handed neutrino masses.

Due to the presence of three Higgs fields, the Higgs potential $V_H(H_S, H_D)$ is more complicated than that of the Standard Model [15, 22, 23]. This potential was first analyzed by Pakvasa and Sugawara [24] who found that in addition to the S_3 symmetry, it has a permutational symmetry S_2': $H_1 \leftrightarrow H_2$, which is not a subgroup of the flavour group S_3. In this communication, we will assume that the vacuum respects the accidental S_2' symmetry of the Higgs potential and that $\langle H_1 \rangle = \langle H_2 \rangle$. With these assumptions, the Yukawa interactions, eqs. (9)-(12) yield mass matrices, for all fermions in the theory, of the general form [14]

$$\mathbf{M} = \begin{pmatrix} \mu_1 + \mu_2 & \mu_2 & \mu_5 \\ \mu_2 & \mu_1 - \mu_2 & \mu_5 \\ \mu_4 & \mu_4 & \mu_3 \end{pmatrix}. \tag{15}$$

$-$	$+$
$H_S,\ \nu_{3R}$	$H_I,\ L_3,\ L_I,\ e_{3R},\ e_{IR},\ \nu_{IR}$

Table 1. Z_2 assignment in the leptonic sector.

The left-handed Majorana neutrinos ν_L naturally acquire their small masses through the see-saw mechanism type I of the form

$$\mathbf{M}_\nu = \mathbf{M}_{\nu_\mathbf{D}} \tilde{\mathbf{M}}_R^{-1} (\mathbf{M}_{\nu_\mathbf{D}})^T, \tag{16}$$

where $\mathbf{M}_{\nu_\mathbf{D}}$ and $\tilde{\mathbf{M}}_R$ denote the Dirac and right handed Majorana neutrino mass matrices, respectively.

In principle, all entries in the mass matrices can be complex since there is no restriction coming from the flavour symmetry S_3. The mass matrices are diagonalized by bi-unitary transformations as

$$U_{iL}^\dagger \mathbf{M}_i U_{iR} = \mathrm{diag}(m_{i1}, m_{i2}, m_{i3}), \quad U_\nu^T \mathbf{M}_\nu U_\nu = \mathrm{diag}(m_{\nu_1}, m_{\nu_2}, m_{\nu_3}). \tag{17}$$

where $i = d, u, e$. The entries in the diagonal matrices may be complex, so the physical masses are their absolute values.

The quark and lepton flavor mixing matrices, V_{PMNS} and V_{CKM}, arise from the mismatch between diagonalization of the mass matrices of u and d type quarks and the diagonalization of the mass matrices of charged leptons and left-handed neutrinos, respectively,

$$V_{CKM} = U_{uL}^\dagger U_{dL}, \quad V_{PMNS} = U_{eL}^\dagger U_\nu K, \tag{18}$$

where K is the diagonal matrix of the Majorana phase factors. Therefore, in order to obtain the unitary matrices appearing in eq. (18) and get predictions for the flavor mixing angles and CP violating phases, we should specify the mass matrices.

3. The mass matrices in the leptonic sector and Z_2 symmetry

A further reduction of the number of parameters in the leptonic sector may be achieved by means of an Abelian Z_2 symmetry. A possible set of charge assignments of Z_2, compatible with the experimental data on masses and mixings in the leptonic sector is given in Table 1.

These Z_2 assignments forbid the following Yukawa couplings, $Y_1^e = Y_3^e = Y_1^\nu = Y_5^\nu = 0$. Therefore, the corresponding entries in the mass matrices vanish, *i.e.*, $\mu_1^e = \mu_3^e = 0$ and $\mu_1^\nu = \mu_5^\nu = 0$.

3.1. The mass matrix of the charged leptons

The mass matrix of the charged leptons takes the form

$$M_e = m_\tau \begin{pmatrix} \tilde{\mu}_2 & \tilde{\mu}_2 & \tilde{\mu}_5 \\ \tilde{\mu}_2 & -\tilde{\mu}_2 & \tilde{\mu}_5 \\ \tilde{\mu}_4 & \tilde{\mu}_4 & 0 \end{pmatrix}. \tag{19}$$

The unitary matrix U_{eL} that enters in the definition of the mixing matrix, V_{PMNS}, is calculated from $U_{eL}^\dagger M_e M_e^\dagger U_{eL} = \mathrm{diag}\left(m_e^2, m_\mu^2, m_\tau^2\right)$, where m_e, m_μ and m_τ are the masses of the charged leptons, and

$$\frac{M_e M_e^\dagger}{m_\tau^2} = \begin{pmatrix} 2|\tilde{\mu}_2|^2 + |\tilde{\mu}_5|^2 & |\tilde{\mu}_5|^2 & 2|\tilde{\mu}_2||\tilde{\mu}_4|e^{-i\delta_e} \\ |\tilde{\mu}_5|^2 & 2|\tilde{\mu}_2|^2 + |\tilde{\mu}_5|^2 & 0 \\ 2|\tilde{\mu}_2||\tilde{\mu}_4|e^{i\delta_e} & 0 & 2\,|\tilde{\mu}_4|^2 \end{pmatrix}. \tag{20}$$

Notice that this matrix has only one non-ignorable phase factor [16].

Once $M_e M_e^\dagger$ has been reparametrized in terms of the charged lepton masses, it is straightforward to compute M_e and U_{eL} also as functions of the charged lepton masses [16]. The resulting expression for M_e, written to order $\left(m_\mu m_e/m_\tau^2\right)^2$ and $x^4 = (m_e/m_\mu)^4$ is

$$M_e \approx m_\tau \begin{pmatrix} \frac{1}{\sqrt{2}}\frac{\tilde{m}_\mu}{\sqrt{1+x^2}} & \frac{1}{\sqrt{2}}\frac{\tilde{m}_\mu}{\sqrt{1+x^2}} & \frac{1}{\sqrt{2}}\sqrt{\frac{1+x^2-\tilde{m}_\mu^2}{1+x^2}} \\ \frac{1}{\sqrt{2}}\frac{\tilde{m}_\mu}{\sqrt{1+x^2}} & -\frac{1}{\sqrt{2}}\frac{\tilde{m}_\mu}{\sqrt{1+x^2}} & \frac{1}{\sqrt{2}}\sqrt{\frac{1+x^2-\tilde{m}_\mu^2}{1+x^2}} \\ \frac{\tilde{m}_e(1+x^2)}{\sqrt{1+x^2-\tilde{m}_\mu^2}}e^{i\delta_e} & \frac{\tilde{m}_e(1+x^2)}{\sqrt{1+x^2-\tilde{m}_\mu^2}}e^{i\delta_e} & 0 \end{pmatrix}. \tag{21}$$

This approximation is numerically exact up to order 10^{-9} in units of the τ mass. Notice that this matrix has no free parameters other than the Dirac phase δ_e.

The unitary matrix U_{eL} that diagonalizes $M_e M_e^\dagger$ and enters in the definition of the neutrino mixing matrix V_{PMNS} may be written as

$$U_{eL} = P_e \mathbf{O}_{eL}, \tag{22}$$

where $P_e = \operatorname{diag}\left(1, 1, e^{i\delta_e}\right)$ and the orthogonal matrix $\mathbf{O}_{eL}$ written to the same order of magnitude as M_e, is

$$\mathbf{O}_{eL} \approx \begin{pmatrix} \frac{1}{\sqrt{2}}x\frac{(1+2\tilde{m}_\mu^2+4x^2+\tilde{m}_\mu^4+2\tilde{m}_e^2)}{\sqrt{1+\tilde{m}_\mu^2+5x^2-\tilde{m}_\mu^4-\tilde{m}_\mu^6+\tilde{m}_e^2+12x^4}} & -\frac{1}{\sqrt{2}}\frac{(1-2\tilde{m}_\mu^2+\tilde{m}_\mu^4-2\tilde{m}_e^2)}{\sqrt{1-4\tilde{m}_\mu^2+x^2+6\tilde{m}_\mu^4-4\tilde{m}_\mu^6-5\tilde{m}_e^2}} & \frac{1}{\sqrt{2}} \\ -\frac{1}{\sqrt{2}}x\frac{(1+4x^2-\tilde{m}_\mu^4-2\tilde{m}_e^2)}{\sqrt{1+\tilde{m}_\mu^2+5x^2-\tilde{m}_\mu^4-\tilde{m}_\mu^6+\tilde{m}_e^2+12x^4}} & \frac{1}{\sqrt{2}}\frac{(1-2\tilde{m}_\mu^2+\tilde{m}_\mu^4)}{\sqrt{1-4\tilde{m}_\mu^2+x^2+6\tilde{m}_\mu^4-4\tilde{m}_\mu^6-5\tilde{m}_e^2}} & \frac{1}{\sqrt{2}} \\ -\frac{\sqrt{1+2x^2-\tilde{m}_\mu^2-\tilde{m}_e^2}(1+\tilde{m}_\mu^2+x^2-2\tilde{m}_e^2)}{\sqrt{1+\tilde{m}_\mu^2+5x^2-\tilde{m}_\mu^4-\tilde{m}_\mu^6+\tilde{m}_e^2+12x^4}} & -x\frac{(1+x^2-\tilde{m}_\mu^2-2\tilde{m}_e^2)\sqrt{1+2x^2-\tilde{m}_\mu^2-\tilde{m}_e^2}}{\sqrt{1-4\tilde{m}_\mu^2+x^2+6\tilde{m}_\mu^4-4\tilde{m}_\mu^6-5\tilde{m}_e^2}} & \frac{\sqrt{1+x^2}\tilde{m}_e\tilde{m}_\mu}{\sqrt{1+x^2-\tilde{m}_\mu^2}} \end{pmatrix}, \tag{23}$$

where, as before, $\tilde{m}_\mu = m_\mu/m_\tau$, $\tilde{m}_e = m_e/m_\tau$ and $x = m_e/m_\mu$.

3.2. The mass matrix of the neutrinos

In the minimal S_3-invariant extension of Standard Model, the Yukawa interactions and the $S_3 \times Z_2$ flavour symmetry yield a mass matrix for the Dirac neutrinos of the form

$$\mathbf{M}_{\nu_\mathrm{D}} = \begin{pmatrix} \mu_2^\nu & \mu_2^\nu & 0 \\ \mu_2^\nu & -\mu_2^\nu & 0 \\ \mu_4^\nu & \mu_4^\nu & \mu_3^\nu \end{pmatrix}. \tag{24}$$

In principle, all entries in the mass matrix $\mathbf{M}_{\nu_\mathrm{D}}$ can be complex, since there is no restriction coming from the $S_3 \times Z_2$ flavour symmetry.

The masses of the left-handed Majorana neutrinos, $\mathbf{M}_\nu$, are generated by the see-saw mechanism type I, $\mathbf{M}_\nu = \mathbf{M}_{\nu_\mathrm{D}}\tilde{\mathbf{M}}_R^{-1}(\mathbf{M}_{\nu_\mathrm{D}})^T$, where $\tilde{\mathbf{M}}_R$ is the mass matrix of the right-handed neutrinos, which we take to be real and diagonal but non-degenerate $\tilde{\mathbf{M}}_R = \operatorname{diag}(M_1, M_2, M_3)$. In previous works it was assumed that the first two right-handed neutrino masses are equal, in which case the resulting reactor mixing angle is different from zero but very small in comparison with the recent experimental data [14, 17]. Hence, in this work we will assumed that $M_1 \neq M_2$.

Then, the mass matrix $\mathbf{M}_\nu$ takes the form

$$\mathbf{M}_\nu = \begin{pmatrix} \frac{2(\mu_2^\nu)^2}{\overline{M}} & \frac{2\lambda(\mu_2^\nu)^2}{\overline{M}} & \frac{2\mu_2^\nu\mu_4^\nu}{\overline{M}} \\ \frac{2\lambda(\mu_2^\nu)^2}{\overline{M}} & \frac{2(\mu_2^\nu)^2}{\overline{M}} & \frac{2\mu_2^\nu\mu_4^\nu\lambda}{\overline{M}} \\ \frac{2\mu_2^\nu\mu_4^\nu}{\overline{M}} & \frac{2\mu_2^\nu\mu_4^\nu\lambda}{\overline{M}} & \frac{2(\mu_4^\nu)^2}{\overline{M}} + \frac{(\mu_3^\nu)^2}{M_3} \end{pmatrix} \tag{25}$$

where

$$\lambda = \frac{1}{2}\left(\frac{M_2 - M_1}{M_1 + M_2}\right) \quad \text{and} \quad \overline{M} = 2\frac{M_1 M_2}{M_2 + M_1} \tag{26}$$

When the first two right-handed neutrino masses are equal, the parameter λ vanishes and we recover the expresion for $\mathbf{M}_\nu$ given in Kubo *et al* [14]. Then, in this model the magnitude of the reactor mixing angle is sensitive to the difference of the values of the first and second masses of the right-handed neutrinos, while the solar and atmospheric mixing angles are almost insensitive to the value of this difference of masses.

As we have assumed the right-handed neutrino mass matrix $\tilde{\mathbf{M}}_R$ to be real, the complex symmetric neutrino mass matrix $\mathbf{M}_\nu$ has only three independent phase factors that come from the parameters μ_2, μ_3 and μ_4. Here, to simplify the analysis we will consider the case when $\arg\{\mu_2^\nu\} = \arg\{\mu_3^\nu\}$ or $2\arg\{\mu_4^\nu\} = \arg\left\{2\frac{(\mu_4^\nu)^2}{\overline{M}} + \frac{(\mu_3^\nu)^2}{M_3}\right\}$. The general case, with three independent phase factors will be considered in detail elsewhere.

In the case considered here, the phase factors may be taken out of $\mathbf{M}_\nu$ as

$$\mathbf{M}_\nu = Q\overline{\mathbf{M}}_\nu Q \tag{27}$$

where

$$Q = \begin{pmatrix} e^{i\phi_2} & 0 & 0 \\ 0 & e^{i\phi_2} & 0 \\ 0 & 0 & e^{i\phi_4} \end{pmatrix} \quad \text{and} \quad \overline{\mathbf{M}}_\nu = \begin{pmatrix} a & d & b \\ d & a & e \\ b & e & c \end{pmatrix} \tag{28}$$

with

$$\phi_2 = \arg\{\mu_2^\nu\},\ \phi_4 = \arg\{\mu_4^\nu\} \tag{29}$$

$$a = \frac{2\,|\mu_2^\nu|^2}{|\overline{M}|},\ b = \frac{2\,|\mu_4^\nu|\,|\mu_2^\nu|}{|\overline{M}|},\ c = \frac{2\,|\mu_4^\nu|^2}{\overline{M}} + \frac{|\mu_3^\nu|^2}{M_3},\ d = \frac{2\,|\lambda|\,|\mu_2^\nu|^2}{|\overline{M}|},\ \text{and}\, e = \frac{2\,|\mu_2^\nu|\,|\mu_4^\nu|\,|\lambda|}{|\overline{M}|} \tag{30}$$

The real symetric matrix $\overline{\mathbf{M}}_\nu$ may be brought to diagonal form by means of a similarity transformation with an orthogonal matrix $\mathbf{O}_\nu$ as $\overline{\mathbf{M}}_\nu = \mathbf{O}_\nu \text{diag}\{m_{\nu_1}, m_{\nu_2}, m_{\nu_3}\}\mathbf{O}_\nu^T$, the columns in $\mathbf{O}_\nu$ are the normalized eigenvectors of $\overline{\mathbf{M}}_\nu$.

In order to compute $\mathbf{O}_\nu$ we notice that the diagonalization of $\overline{\mathbf{M}}_\nu$ is equivalent to the diagonalization of a mass matrix $\widehat{\mathbf{M}}$ with two texture zeroes. First define a new mass matrix

$\overline{\mathbf{M}}_\nu{}'$ obtained from $\overline{\mathbf{M}}_\nu$ by a $\frac{\pi}{4}$ rotation

$$\mathcal{U}_{\frac{\pi}{4}} = \begin{pmatrix} \frac{1}{\sqrt{2}} & 0 & \frac{1}{\sqrt{2}} \\ -\frac{1}{\sqrt{2}} & 0 & \frac{1}{\sqrt{2}} \\ 0 & 1 & 0 \end{pmatrix}, \tag{31}$$

through the similarity transformation $\overline{\mathbf{M}}_\nu{}' = \mathcal{U}_{\frac{\pi}{4}}^\dagger \, \overline{\mathbf{M}}_\nu \mathcal{U}_{\frac{\pi}{4}}$.Then, the matrix $\overline{\mathbf{M}}_\nu{}'$ can be written in the following form:

$$\overline{\mathbf{M}}_\nu{}' = \mu_0 \mathbb{I}_{3\times3} + \widehat{\mathbf{M}} \tag{32}$$

where $\mathbb{I}_{3\times3}$ is the identity matrix,

$$\mu_0 = a - d = \frac{2\left|\mu_2^\nu\right|^2}{\left|\overline{M}\right|}(1 - |\lambda|) \qquad \text{and} \qquad \widehat{\mathbf{M}} = \begin{pmatrix} 0 & A & 0 \\ A & B & C \\ 0 & C & D \end{pmatrix} \tag{33}$$

with $A = \frac{b-e}{\sqrt{2}}$, $B = c + d - a$, $C = \frac{b+e}{\sqrt{2}}$ and $D = 2d$. As mentioned before the diagonalization of $\overline{\mathbf{M}}_\nu$ is reduced to the diagonalization of the real symmetric matrix $\widehat{\mathbf{M}}$, which is a matrix with two texture zeroes of class I [12]. In the literature, these similarity transformations are also known as weak basis transformations, since they leave invariant the gauge currents [25].

As in the case of the charged leptons, the matrices $\mathbf{M}_\nu$ and U_ν can be reparametrized in terms of the neutrino masses. For this we use the information that we already have about the diagonalization of a matrix with two texture zeroes of class I [12, 13, 21]. Then, the mass matrix $\mathbf{M}_\nu$ for a normal [inverted] hierarchy in the mass spectrum takes the form

$$\mathbf{M}_\nu^{\mathbf{N[I]}} = \begin{pmatrix} \mu_0 + d & d & \frac{1}{\sqrt{2}}\left(C^{N[I]} + A^{N[I]}\right) \\ d & \mu_0 + d & \frac{1}{\sqrt{2}}\left(C^{N[I]} - A^{N[I]}\right) \\ \frac{1}{\sqrt{2}}\left(C^{N[I]} + A^{N[I]}\right) & \frac{1}{\sqrt{2}}\left(C^{N[I]} - A^{N[I]}\right) & m_{\nu_1} + m_{\nu_2} + m_{\nu_3} - 2(\mu_0 + d) \end{pmatrix} \tag{34}$$

where

$$A^{N[I]} = \sqrt{\frac{(m_{\nu_2} - \mu_0)\left(m_{\nu_{3[1]}} - \mu_0\right)\left(\mu_0 - m_{\nu_{1[3]}}\right)}{2d}}, \tag{35}$$

$$C^{N[I]} = \sqrt{\frac{(2d + \mu_0 - m_{\nu_1})\left(2d + \mu_0 - m_{\nu_{2[3]}}\right)\left(m_{\nu_{3[2]}} - \mu_0 - 2d\right)}{2d}}. \tag{36}$$

The values allowed for the parameters μ_0 and $2d + \mu_0$ are in the following ranges:

$$m_{\nu_{2[1]}} > \mu_0 > m_{\nu_{1[3]}} \quad \text{and} m_{\nu_{3[2]}} > 2d + \mu_0 > m_{\nu_{2[1]}} \tag{37}$$

Now, the unitary matrix U_ν takes the following form:

$$U_\nu = P_\nu \mathcal{U}_{\frac{\pi}{4}} \mathbb{O}^{N[I]}, \tag{38}$$

where $\mathcal{U}_{\frac{\pi}{4}}$ is given in eq.(31), $P_\nu = \text{diag}\left\{1, 1, e^{i\delta_\nu}\right\}$ with $\delta_\nu = \phi_2 - \phi_4$, and $\mathbb{O}^{N[I]}$ is the matriz that diagonalizes $\widehat{\mathbf{M}}$ whose elements are

$$
\begin{aligned}
O_{11}^{N[I]} &= \sqrt{\frac{[-1](m_{\nu_3}-\mu_0)(m_{\nu_2}-\mu_0)f_1}{\mathcal{D}_1^{N[I]}}}, & O_{12}^{N[I]} &= \sqrt{\frac{\left(m_{\nu_{3[1]}}-\mu_0\right)\left(\mu_0-m_{\nu_{1[3]}}\right)f_2^{N[I]}}{\mathcal{D}_2^{N[I]}}}, \\
O_{13}^{N[I]} &= -\sqrt{\frac{[-1](\mu_0-m_{\nu_1})(m_{\nu_2}-\mu_0)f_3^{N[I]}}{\mathcal{D}_3^{N[I]}}}, & O_{21}^{N[I]} &= \sqrt{\frac{[-1]2d(\mu_0-m_{\nu_1})f_1}{\mathcal{D}_1^{N[I]}}}, \\
O_{22}^{N[I]} &= \sqrt{\frac{2d(m_{\nu_2}-\mu_0)f_2^{N[I]}}{\mathcal{D}_2^{N[I]}}}, & O_{23}^{N[I]} &= \sqrt{\frac{[-1]2d(m_{\nu_3}-\mu_0)f_3^{N[I]}}{\mathcal{D}_3^{N[I]}}}, \\
O_{31}^{N[I]} &= -\sqrt{\frac{[-1](\mu_0-m_{\nu_1})f_2^{N[I]}f_3^{N[I]}}{\mathcal{D}_1^{N[I]}}}, & O_{32}^{N[I]} &= \sqrt{\frac{(m_{\nu_2}-\mu_0)f_1f_3^{N[I]}}{\mathcal{D}_1^{N[I]}}}, \quad O_{33}^{N[I]} = -\sqrt{\frac{(m_{\nu_3}-\mu_0)f_1f_2^{N[I]}}{\mathcal{D}_1^{N[I]}}},
\end{aligned}
\tag{39}
$$

where

$$
\begin{aligned}
\mathcal{D}_1^{N[I]} &= 2d\,(m_{\nu_2} - m_{\nu_1})\left(m_{\nu_{3[1]}} - m_{\nu_{1[3]}}\right) & \mathcal{D}_2^{N[I]} &= 2d\,(m_{\nu_2} - m_{\nu_1})\left(m_{\nu_{3[2]}} - m_{\nu_{2[3]}}\right) \\
\mathcal{D}_3^{N[I]} &= 2d\left(m_{\nu_{3[1]}} - m_{\nu_{1[3]}}\right)\left(m_{\nu_{3[2]}} - m_{\nu_{2[3]}}\right) & f_1 &= (2d + \mu_0 - m_{\nu_1}) \\
f_2^{N[I]} &= [-1]\,(2d + \mu_0 - m_{\nu_2}) & f_3^{N[I]} &= [-1]\,(m_{\nu_3} - \mu_0 - 2d)
\end{aligned}
\tag{40}
$$

The superscripts N and I denote the normal and inverted hierarchies respectively.

It is well know [21] that in the hadronic sector the masses of quarks may be obtained from a matrix with two texture zeroes which succesfully reproduces the strong mass hierarchy of up and down type quarks. Also, the numerical values of the quark mixing angles determined in this framework are in good agreement with the experimental data [13]. Additionally, in a unified treatment in which the mass matrices of all fermions have a similar form with two texture zeroes of class I and a normal hierarchy, the numerical values obtained for masses and mixing of the neutrinos are in very good agreement with all available experimental data [13, 10]. Therefore, it is to be expected that the mixing matrix V_{PMNS} that is obtained from the mass matrices $\mathbf{M}_\nu$ and M_e, eqs. (21)-(25) will also reproduce the current experimental data on masses and mixings in the leptonic sector of the theory.

4. The neutrino mixing matrix

The neutrino mixing matrix V_{PMNS}, is the product $U_{eL}^\dagger U_\nu K$, where K is the diagonal matrix of the Majorana phase factors, defined by $K = diag(1, e^{i\alpha}, e^{i\beta})$. Now, we obtain the theoretical expression of the elements of the lepton mixing matrix, V_{PMNS}^{th}. This expression has the following form:

$$
V_{PMNS}^{th} = \begin{pmatrix} V_{e1}^{th} & V_{e2}^{th}e^{i\alpha} & V_{e3}^{th}e^{i\beta} \\ V_{\mu1}^{th} & V_{\mu2}^{th}e^{i\alpha} & V_{\mu3}^{th}e^{i\beta} \\ V_{\tau1}^{th} & V_{\tau2}^{th}e^{i\alpha} & V_{\tau3}^{th}e^{i\beta} \end{pmatrix}
\tag{41}
$$

where

$$V_{e1}^{th} = \frac{\tilde{m}_e}{\tilde{m}_\mu} O_{11}^{N[I]} - O_{21}^{N[I]} e^{i\delta_l}, \quad V_{e2}^{th} = \frac{\tilde{m}_e}{\tilde{m}_\mu} O_{12}^{N[I]} - O_{22}^{N[I]} e^{i\delta_l}, \quad V_{\tau 1}^{th} = O_{31}^{N[I]},$$
$$V_{e3}^{th} = \frac{\tilde{m}_e}{\tilde{m}_\mu} O_{13}^{N[I]} - O_{23}^{N[I]} e^{i\delta_l}, \quad V_{\mu 1}^{th} = -O_{11}^{N[I]} - \frac{\tilde{m}_e}{\tilde{m}_\mu} O_{21}^{N[I]} e^{i\delta_l}, \quad V_{\tau 2}^{th} = O_{32}^{N[I]}, \tag{42}$$
$$V_{\mu 2}^{th} = -O_{12}^{N[I]} - \frac{\tilde{m}_e}{\tilde{m}_\mu} O_{22}^{N[I]} e^{i\delta_l}, \quad V_{\mu 3}^{th} = -O_{13}^{N[I]} - \frac{\tilde{m}_e}{\tilde{m}_\mu} O_{23}^{N[I]} e^{i\delta_l}, \quad V_{\tau 3}^{th} = O_{33}^{N[I]}$$

with $\delta_l = \delta_\nu - \delta_e$ and the elements $O_{ij}^{N[I]}$ $(i, j = 1, 2, 3)$ in the eqs. (42) are given in the eq. (39).

4.1. *The Mixing Angles*

The theoretical expression for the lepton mixing angles as functions of the lepton mass ratios and one phase are:

$$\sin^2 \theta_{12}^l = \frac{\left(\frac{\tilde{m}_e}{\tilde{m}_\mu}\right)^2 \left(O_{12}^{N[I]}\right)^2 + \left(O_{22}^{N[I]}\right)^2 - 2\frac{\tilde{m}_e}{\tilde{m}_\mu} O_{12}^{N[I]} O_{22}^{N[I]} \cos\delta_l}{1 - \left(\frac{\tilde{m}_e}{\tilde{m}_\mu}\right)^2 \left(O_{13}^{N[I]}\right)^2 - \left(O_{23}^{N[I]}\right)^2 + 2\frac{\tilde{m}_e}{\tilde{m}_\mu} O_{13}^{N[I]} O_{23}^{N[I]} \cos\delta_l} \tag{43}$$

$$\sin^2 \theta_{23}^l = \frac{\left(O_{13}^{N[I]}\right)^2 + \left(\frac{\tilde{m}_e}{\tilde{m}_\mu}\right)^2 \left(O_{23}^{N[I]}\right)^2 + 2\frac{\tilde{m}_e}{\tilde{m}_\mu} O_{13}^{N[I]} O_{23}^{N[I]} \cos\delta_l}{1 - \left(\frac{\tilde{m}_e}{\tilde{m}_\mu}\right)^2 \left(O_{13}^{N[I]}\right)^2 - \left(O_{23}^{N[I]}\right)^2 + 2\frac{\tilde{m}_e}{\tilde{m}_\mu} O_{13}^{N[I]} O_{23}^{N[I]} \cos\delta_l} \tag{44}$$

$$\sin^2 \theta_{13}^l = \left(\frac{\tilde{m}_e}{\tilde{m}_\mu}\right)^2 \left(O_{13}^{N[I]}\right)^2 + \left(O_{23}^{N[I]}\right)^2 - 2\frac{\tilde{m}_e}{\tilde{m}_\mu} O_{13}^{N[I]} O_{23}^{N[I]} \cos\delta_l \tag{45}$$

where, as before, $\tilde{m}_\mu = m_\mu/m_\tau$ and $\tilde{m}_e = m_e/m_\tau$.

In a first, preliminary analysis for the mixing angle θ_{13}^l and for an inverted neutrino mass hierarchy $(m_{\nu_2} > m_{\nu_1} > m_{\nu_3})$ the eq. (45) takes the form:

$$\sin^2 \theta_{13}^l \approx \frac{(\mu_0 + 2d - m_{\nu_3})(\mu_0 - m_{\nu_3})}{(m_{\nu_1} - m_{\nu_3})(m_{\nu_2} - m_{\nu_3})}. \tag{46}$$

Now, with the following values for neutrino masses $m_{\nu_2} = 0.056\,eV$, $m_{\nu_1} = 0.053\,eV$ and $m_{\nu_3} = 0.048\,eV$, and the parameters values $\delta_l = \pi/2$, $\mu_0 = 0.049\,eV$ and $d = 8 \times 10^{-5}\,eV$ we get $\sin^2 \theta_{13}^l \approx 0.029 \longrightarrow \theta_{13}^l \approx 9.8^\circ$.

5. Conclusions

In the minimal S_3-invariant extension of the Standard Model [14] a well defined structure of the Yukawa couplings is obtained, which permits the calculations of mass and mixing matrices for quark and leptons in a unified way. A further reduction of redundant parameters is achieved in the leptonic sector by introducing a Z_2 symmetry. The flavour symmetry group $S_3 \times Z_2$ relates the neutrino mass spectrum and mixings. This allowed us to compute the neutrino mixing matrix, V_{PMNS}, explicitly in terms of the masses of the charged leptons and neutrinos and one phase δ_l. In this model, the mass matrices of all fermions are brought to the same generic form with two texture zeroes by means of a similarity transformation. Then, we computed the neutrino mixing matrix $V_{_{PMNS}}$ which has three CP-violating phases, namely, one Dirac phase $\delta_l = \delta_\nu - \delta_e$ and two Majorana phases α ans β which are funtions of the charged leptons and neutrino masses. The neutrino mixing angles are given as funtions of the charged leptons and neutrino masses. In this way, the numerical value of the reactor mixing angle, θ_{13}, is related to the numerical values of the neutrinos masses, which in this analysis have an inverted mass hierarchy with the values $|m_{\nu_2}| = 0.056\ eV$, $|m_{\nu_1}| = 0.053\ eV$ and $|m_{\nu_3}| = 0.048\ eV$. The numerical value obtained for θ_{13} is $\theta_{13}^l \approx 9.8^\circ$ in agreement with the latest analysis of the experimental data on neutrino oscillations and mixings [2, 3].

6. Acknowledgements
This work was partially supported by DGAPA-UNAM under contract PAPIIT-IN113712-3.

References
[1] Jung C K, McGrew C, Kajita T and Mann T 2001 *Ann. Rev. Nucl. Part. Sci.* **51** 451–488. Mohapatra R N and Smirnov A Y 2006 *Ann. Rev. Nucl. Part. Sci.* **56** 569–628 (*Preprint* hep-ph/0603118). Altmann M *et al.* (GNO) 2005 *Phys. Lett.* **B616** 174–190 (*Preprint* hep-ex/0504037). Smy M B *et al.* (Super-Kamiokande) 2004 *Phys. Rev.* **D69** 011104 (*Preprint* hep-ex/0309011). Ahmad Q R *et al.* (SNO) 2002 *Phys. Rev. Lett.* **89** 011302 (*Preprint* nucl-ex/0204009). Ahmad Q R *et al.* (SNO) 2002 *Phys. Rev. Lett.* **89** 011301 (*Preprint* nucl-ex/0204008). Aharmim B *et al.* (SNO) 2005 *Phys. Rev.* **C72** 055502 (*Preprint* nucl-ex/0502021). Fukuda S *et al.* (Super-Kamiokande) 2002 *Phys. Lett.* **B539** 179–187 (*Preprint* hep-ex/0205075). Ashie Y *et al.* (Super-Kamiokande) 2005 *Phys. Rev.* **D71** 112005 (*Preprint* hep-ex/0501064). Bemporad C, Gratta G and Vogel P 2002 *Rev. Mod. Phys.* **74** 297 (*Preprint* hep-ph/0107277). Araki T *et al.* (KamLAND) 2005 *Phys. Rev. Lett.* **94** 081801 (*Preprint* hep-ex/0406035). Maltoni M, Schwetz T, Tortola M A and Valle J W F 2004 *New J. Phys.* **6** 122 (*Preprint* hep-ph/0405172). Schwetz T 2005 *Acta Phys. Polon.* **B36** 3203–3214 (*Preprint* hep-ph/0510331). Gonzalez-Garcia M C and Maltoni M 2008 *Phys. Rept.* **460** 1–129 (*Preprint* 0704.1800). Apollonio M *et al.* (CHOOZ) 2003 *Eur. Phys. J.* **C27** 331–374 (*Preprint* hep-ex/0301017). K and Eitel 2005 *Nuclear Physics B - Proceedings Supplements* **143** 197 – 204. Elliott S R and Engel J 2004 *J. Phys.* **G30** R183 (*Preprint* hep-ph/0405078). Seljak U, Slosar A and McDonald P 2006 *JCAP* **0610** 014 (*Preprint* astro-ph/0604335). Elgaroy O and Lahav O 2005 *New J. Phys.* **7** 61 (*Preprint* hep-ph/0412075). Lesgourgues J and Pastor S 2006 *Phys. Rept.* **429** 307–379 (*Preprint* astro-ph/0603494).
[2] Abe K *et al.* (T2K Collaboration) 2011 *Phys. Rev. Lett.* **107**(4) 041801
[3] Abe Y *et al.* (DOUBLE-CHOOZ) 2011 (*Preprint* 1112.6353)
[4] Schwetz T, Tortola M and Valle J W F 2011 *New J. Phys.* **13** 109401 (*Preprint* 1108.1376)
[5] Gonzalez-Garcia M C, Maltoni M and Salvado J 2010 *JHEP* **04** 056 (*Preprint* 1001.4524). Gonzalez-Garcia M 2011 *Physics of Particles and Nuclei* **42**(4) 577–589
[6] Schwetz T, Tortola M and Valle J W F 2011 *New J. Phys.* **13** 063004 (*Preprint* 1103.0734)
[7] An, F.P. and others *Phys. Rev. Lett.* **108** 171803 (*Preprint* 1203.1669)
[8] Ahn, J.K. and others (*Preprint* 1204.0626)
[9] Fritzsch H and Xing Z z 2000 *Prog. Part. Nucl. Phys.* **45** 1–81 (*Preprint* hep-ph/9912358)
[10] Xing Z z and Zhang H 2003 *Phys. Lett.* **B569** 30–40 (*Preprint* hep-ph/0304234)
[11] Ishimori H *et al.* 2010 *Prog. Theor. Phys. Suppl.* **183** 1–163 (*Preprint* 1003.3552). Altarelli G and Feruglio F 2010 *Rev. Mod. Phys.* **82**(3) 2701–2729
[12] Gonzalez Canales F and Mondragon A 2011 *J. Phys. Conf. Ser.* **287** 012015 (*Preprint* 1101.3807)
[13] Barranco J, Gonzalez Canales F and Mondragon A 2010 *Phys. Rev.* **D82** 073010 (*Preprint* 1004.3781)
[14] Kubo J, Mondragon A, Mondragon M and Rodriguez-Jauregui E 2003 *Prog. Theor. Phys.* **109** 795–807 (*Preprint* hep-ph/0302196)
[15] Kubo J *et al.* 2005 *J. Phys. Conf. Ser.* **18** 380–384
[16] Felix O, Mondragon A, Mondragon M and Peinado E 2007 *AIP Conf. Proc.* **917** 383–389 (*Preprint* hep-ph/0610061)
[17] Mondragon A, Mondragon M and Peinado E 2007 *Phys. Rev.* **D76** 076003 (*Preprint* 0706.0354)
[18] Mondragon A, Mondragon M and Peinado E 2008 *J. Phys.* **A41** 304035 (*Preprint* 0712.1799)
[19] Mondragon A, Mondragon M and Peinado E 2008 *AIP Conf. Proc.* **1026** 164–169 (*Preprint* 0712.2488)
[20] Mondragon A, Mondragon M and Peinado E 2008 *Rev. Mex. Fis.* **S54N3** 81–91 (*Preprint* 0805.3507)
[21] Mondragon A and Rodriguez-Jauregui E 1999 *Phys. Rev.* **D59** 093009 (*Preprint* hep-ph/9807214). Mondragon A and Rodriguez-Jauregui E 2000 *Phys. Rev.* **D61** 113002 (*Preprint* hep-ph/9906429).
[22] Emmanuel-Costa D, Felix-Beltran O, Mondragon M and Rodriguez-Jauregui E 2007 *AIP Conf. Proc.* **917** 390–393
[23] Teshima T 2012 (*Preprint* 1202.4528)
[24] Pakvasa S and Sugawara H 1978 *Phys. Lett.* **B73** 61
[25] Branco G C, Emmanuel-Costa D, Gonzalez Felipe R and Serodio H 2009 *Phys. Lett.* **B670** 340–349 (*Preprint* 0711.1613)
[26] Gonzalez Canales F, Mondragon A *et al.* (*Preprint* Work in progress)
[27] Fritzsch H and Xing Z Z 1997 *Phys. Lett.* **B413** 396–404 (*Preprint* hep-ph/9707215)

Journal of Physics: Conference Series **387** (2012) 012009 doi:10.1088/1742-6596/387/1/012009

Novel Bose-Einstein Interference in the Passage of a Fast Particle in a Dense Medium[1]

C Y Wong
Physics Division, Oak Ridge National Laboratory, Oak Ridge, TN 37831
E-mail: wongc@ornl.gov

Abstract.
When an energetic particle collides coherently with many medium particles at high energies, the Bose-Einstein symmetry with respect to the interchange of the exchanged virtual bosons leads to a destructive interference of the Feynman amplitudes in most regions of the phase space but a constructive interference in some other regions of the phase space. As a consequence, the recoiling medium particles have a tendency to come out collectively along the direction of the incident fast particle, each carrying a substantial fraction of the incident longitudinal momentum. Such an interference appearing as collective recoils of scatterers along the incident particle direction may have been observed in angular correlations of hadrons associated with a high-p_T trigger in high-energy AuAu collisions at RHIC.

1. Introduction

In the collision of an energetic particle $p=(p_0, \boldsymbol{p})$ with n medium particles, how dense must the medium be for the multiple collisions to become a single coherent $(1+n)$-body collision, instead of a sequence of n incoherent 2-body collisions? It is instructive to find out the conditions on the medium density and the energy of the incident particle that determine whether the set of multiple collisions are coherent or incoherent [1]. For such a purpose, we consider a binary collision between the incident fast particle p and a medium scatterer a_i with the exchange of a boson, in the medium center-of-momentum frame. The longitudinal momentum transfer q_z for the binary collision can be obtained from the transverse momentum transfer q_T by $q_z \sim q_T^2/2p_0$. The longitudinal momentum transfer is associated with a longitudinal coherence length $\Delta z_{\rm coh} \sim \hbar/q_z \sim 2\hbar p_0/q_T^2$ that specifies the uncertainties in the longitudinal locations at which the virtual boson is exchanged between the fast particle and the scatterer.

The nature of the multiple scattering process can be inferred by comparing the longitudinal coherence length $\Delta z_{\rm coh}$ with the mean free path λ of the jet in the dense medium that depends not only on the density of the medium but also on the binary collision cross section. If $\Delta z_{\rm coh} \ll \lambda$, then a single binary collision is well completed before another binary collision begins, and the multiple collision process consists of a sequence of n incoherent two-body collisions.

If $\Delta z_{\rm coh} \gg \lambda$, then a single binary collision is not completed before another one begins, and the multiple collision process consists of a set of coherent collisions as a single $(1+n)$-body collision. For a set of initial and final states in such a coherent $(1+n)$-body collision, there are

[1] Based in part on an invited talk presented at the 35th Symposium on Nuclear Physics, Cocoyoc, Mexico, January 3, 2012.

$n!$ different trajectories in the sequence of collisions along $\Delta z_{\rm coh}$ at which various virtual bosons are exchanged. By Bose-Einstein symmetry, the total Feynman amplitude is then the sum of the amplitudes for all possible interchanges of the exchanged virtual bosons. The summation of these Feynman amplitudes and the accompanying interference constitute the Bose-Einstein interference in the passage of the fast particle in the dense medium.

There is another important effect that accompanies the coherent (1+n)-body collision and changes the nature of the collision process. A sequence of n incoherent two-body collisions contains only $2n$ degrees of freedom, which can be chosen to be the transverse momentum transfers $\{\boldsymbol{q}_{1T}, \boldsymbol{q}_{2T}, \boldsymbol{q}_{3T}, ..., \boldsymbol{q}_{nT}\}$. The longitudinal momentum transfer q_z after each individual jet-parton collision is then a dependent variable, depending on the corresponding transverse momentum transfer as $q_{iz} \sim |\boldsymbol{q}_{iT}|^2/2p_0$. In contrast, the coherent (1+n)-body collision links the incident particle with n scatterers as a single collisional unit. There are $3n-1$ degrees of freedom in this coherent (1+n)-body collision, after the $3(n+1)$ degrees of freedom are reduced by the constraints of the conservation of energy and momentum. The set of longitudinal momentum transfers, $\{q_{iz}, i = 1, 2, ..., n\}$, can also be independent variables with their own probability distribution functions. The longitudinal momentum of the incident fast particle can flow to the n scatterers, which can then carry a substantial fraction of the longitudinal momentum of the incident fast particle. Thus, the degrees of freedom increases from $2n$ for incoherent collisions to $3n-1$ for a coherent collision, and this increase changes the character of the collision process.

It is of interest to inquire whether there are fast particles energetic enough, scatterer media dense enough, and binary cross section large enough, for coherent collisions to occur. In high energy central collisions between heavy nuclei such as those at RHIC and LHC, both jets (mini-jets) and a dense medium are produced after each collision. The jets will collide with partons in the dense medium, and these collisions may satisfy the condition for coherent collisions. The longitudinal coherent length $\Delta z_{\rm coh}$ is of order 25 fm, for a typical transverse momentum transfer of $q_T{\sim}0.4$ GeV/c from a jet of momentum $p_0{\sim}10$ GeV/c to a medium parton in a binary collision at RHIC [2]. The longitudinal coherent length $\Delta z_{\rm coh}$ is much greater than the radius R of a large nucleus. On the other hand, the away side jet is quenched by the dense medium in the most central AuAu collisions at RHIC and LHC [3, 4, 5], and the near-side jet collides with about 4-6 medium partons [6]-[13]. Therefore, one can infer that the mean-free path λ for the collision of the jet with medium partons is much smaller than the nuclear radius R. Hence, $\Delta z_{\rm coh} \gg R \gg \lambda$ in high-energy central nuclear collisions at RHIC and LHC, and the multiple collision process of a jet with medium partons along its path constitutes a set of coherent collision.

2. Bose-Einstein Interference of Feynman Amplitudes

As an example of the interference of Feynman amplitudes in a coherent collision, we consider

$$p + a_1 + a_2 \to p' + a_1' + a_2', \tag{1}$$

the collisions of an energetic fermion p with two fermion scatterers a_1 and a_2 of rest mass m in the Abelian gauge theory. The Feynman amplitude for diagram 1(a) is given by [14]

$$M_a = -g^4 \bar u(\boldsymbol{p}')\gamma_\nu \frac{1}{\not p - \not q_1 - m + i\epsilon'}\gamma_\mu u(\boldsymbol{p})\frac{1}{q_2^2}\bar u(\boldsymbol{a}_2')\gamma_\nu u(\boldsymbol{a}_2)\frac{1}{q_1^2}\bar u(\boldsymbol{a}_1')\gamma_\mu u(\boldsymbol{a}_1). \tag{2}$$

If the spatial separation between the scatterers is so large that $\lambda \gg \Delta z_{\rm coh}$, then diagram 1(a) can be cut into two disjoint pieces and the collision process consists effectively of a set of two incoherent two-body collisions.

On the other hand, if the collision process is characterized by $\lambda \ll \Delta z_{\rm coh}$, it comprises a set of coherent (1+2)-body collision. There is an additional Feynman amplitude M_b for diagram

Journal of Physics: Conference Series **387** (2012) 012009 doi:10.1088/1742-6596/387/1/012009

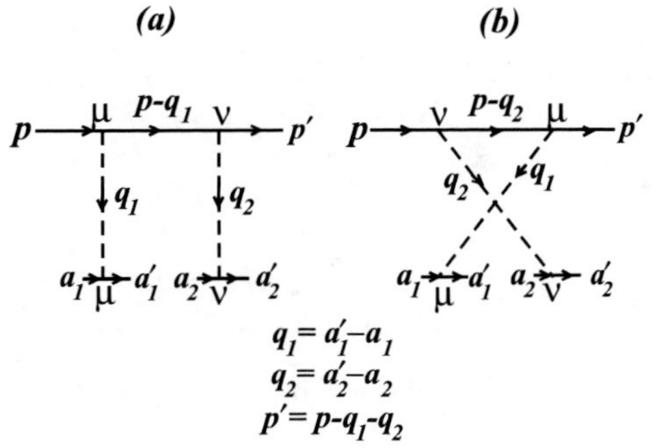

Figure 1. Feynman diagrams for the collision of a fast fermion p with medium fermions a_1, a_2, with the emission and absorption of virtual bosons of momenta q_1 and q_2.

1(b), obtained by making a symmetrized permutation of the bosons in diagram 1(a),

$$M_b \quad = \quad -g^4\bar{u}(\boldsymbol{p}')\gamma_\mu\frac{1}{\not{p}-\not{q}_2-m+i\epsilon'}\gamma_\nu u(\boldsymbol{p})\frac{1}{q_2^2}\bar{u}(\boldsymbol{a}_2')\gamma_\nu u(\boldsymbol{a}_2)\frac{1}{q_1^2}\bar{u}(\boldsymbol{a}_1{}')\gamma_\mu u(\boldsymbol{a}_1). \tag{3}$$

The trajectories for diagram 1(a) and 1(b) are both possible paths in a coherent collision, leading from the a set of initial states to a set of final states. By Bose-Einstein symmetry, the total amplitude M for coherent collisions is the symmetrized sum of M_a and M_b.

We consider the high-energy limit and assume the conservation of helicity with

$$\{p_0, |\boldsymbol{p}|, p_0', |\boldsymbol{p}'|\} \gg \{|\boldsymbol{a}_i|, q_0, |\boldsymbol{q}_i|\} \gg m, \text{ for } i=1,2. \tag{4}$$

In this limit, we have approximately [14]

$$\bar{u}(\boldsymbol{a}')\gamma_\mu u(\boldsymbol{a}) \sim \sqrt{\frac{a_0+m}{a_0'+m}}\frac{a_\mu'}{2m} + \sqrt{\frac{a_0'+m}{a_0+m}}\frac{a_\mu}{2m} \equiv \frac{\tilde{a}_\mu}{m}, \tag{5}$$

$$\frac{1}{\not{p}-\not{q}_1-m+i\epsilon'}\gamma_\nu u(p) \sim -\frac{\not{p}-\not{q}_1+m}{2p\cdot q_1 - i\epsilon}\gamma_\nu u(p), \tag{6}$$

$$\bar{u}(p)\gamma_\nu(\not{p}-\not{q}_1+m)\gamma_\mu u(p) \sim \frac{2p_\nu p_\mu}{m}, \tag{7}$$

where ϵ is a small positive quantity. We shall be interested in the case in which the fermion p' after the collision remains on the mass shell. The mass shell condition can be expressed as

$$(p-q_1-q_2)^2 - m^2 \sim -2p\cdot q_1 - 2p\cdot q_2 \sim 0. \tag{8}$$

The symmetrized sum of the Feynman amplitudes M_a and M_b in the high-energy limit is

$$M \sim \frac{g^4}{2m}\frac{2p\cdot\tilde{a}_1}{m}\frac{2p\cdot\tilde{a}_2}{m}\frac{1}{q_2^2q_1^2}\left(\frac{1}{2p\cdot q_1 - i\epsilon} + \frac{1}{2p\cdot q_2 - i\epsilon}\right). \tag{9}$$

Note that the amplitudes M_a and M_b correlate with each other because of the mass-shell condition (8). The real parts of the amplitudes destructively cancel, and the imaginary parts interfere and add constructively, to result in sharp distributions at $p\cdot q_1 \sim 0$ and $p\cdot q_2 \sim 0$,

$$M \sim \frac{g^4}{2m}\frac{2p\cdot\tilde{a}_1}{m}\frac{2p\cdot\tilde{a}_2}{m}\frac{1}{q_2^2q_1^2}\Big\{i\pi\Delta(2p\cdot q_1) + i\pi\Delta(2p\cdot q_2)\Big\}, \quad \Delta(2p\cdot q_1) = \frac{1}{\pi}\frac{\epsilon}{(2p\cdot q_1)^2+\epsilon^2}, \tag{10}$$

where the function $\Delta(2p \cdot q_1)$ approaches the Dirac delta function $\delta(2p \cdot q_1)$ in the limit $\epsilon \to 0$.

Generalizing to the case of the coherent collision of a fast fermion with n fermion scatterers, the total Feynman amplitude is

$$M = \frac{g^{2n}}{2m} \left\{ \prod_{i=1}^{n} \frac{2p \cdot \tilde{a}_i}{m q_i^2} \right\} \mathcal{M}(q_1, q_2, ..., q_n), \tag{11}$$

where $\mathcal{M}(q_1, q_2, ..., q_n)$ is the sum of $n!$ amplitudes involving symmetric permutations of the exchanged bosons given by

$$\mathcal{M}(q_1, q_2, ..., q_n) = \prod_{j=1}^{n-1} \frac{1}{\sum_{i=1}^{j} 2p \cdot q_i - i\epsilon} + \text{symmetric permutations.} \tag{12}$$

The above sum involves extensive cancellations. Remarkably, it can be shown that this sum of $n!$ permutations turns out to be a product of sharp distributions centered at $2p \cdot q_i \sim 0$ [14],

$$\mathcal{M}(q_1, q_2, ..., q_n) \Delta(\sum_{i=1}^{n} 2p \cdot q_i) = (2\pi i)^{n-1} \prod_{i=1}^{n} \Delta(2p \cdot q_i), \tag{13}$$

which indicates that in coherent collisions, there is a destructive interference of the Feynman amplitudes in most regions of the phase space but a constructive interference in some other regions of the the phase space, leading to sharp distributions at $2p \cdot q_i \sim 0$.

Equations (11)-(13) for the total Feynman amplitude as a product of delta functions of $2p \cdot q_i$ for a coherent collision are similar to previous results obtained for the emission of many real photons or gluons in bremsstrahlung, and for the sum of ladder and cross-ladder amplitudes in the collision of two fermions [14]-[21].

3. Consequences of the BE Interference on the Recoils of Fermion Scatterers

The differential cross section for the coherent collision $p + a_1 + ... + a_n \to p' + a'_1 + ... + a'_n$ is

$$d^n\sigma = \frac{1}{f_{j1} f_{j2} ... f_{jn} T_2 T_3 ... T_n} |M|^2 (2\pi)^4 \delta^4(p' + \sum_{i=1}^{n} q_i - p) \frac{d\boldsymbol{p}'_i 2m}{(2\pi)^3 2p'_{i0}} \prod_{i=1}^{n} \frac{d^4 a'_i 2m}{(2\pi)^3 2a'_{i0}} D_i(a'_i), \tag{14}$$

where f_{ji} is the dimensionless flux factor between the fast particle p and scatterer a_i,

$$f_{ji} = \frac{\sqrt{p \cdot a_i - m\, m_{iT}}}{m_i\, m}, \tag{15}$$

and $\{T_i, i = 1, ..(n-1)\}$ are the collision times for the intermediate state of the fast particle p to collide with the scatterers after the ith collision, and $m_{iT} = \sqrt{m_i^2 + \boldsymbol{a}_T^2}$. The function D_i describes the mass-shell conditions of medium scatterers after collision,

$$D_i(a'_i) = \frac{\Gamma_i/2}{\pi\{[(a'_{i0} - \sqrt{(\boldsymbol{a}'_i - \boldsymbol{A})^2 + (m_i + S)^2} + A_0]^2 + \Gamma_i^2/4\}}, \tag{16}$$

where $A = \{\boldsymbol{A}, A_0\}$ and S are the vector and scalar mean fields experienced by the medium scatterer a'_i after collision, respectively, and Γ_i is the width of a'_i. As the mean fields and scatterer widths increase with density and are presumably quite large and dominant for a dense medium, we approximately represent D_i as an average constant. We change variables from a_i' to $q_i = a_i' - a_i$, and integrate over $\boldsymbol{p}'$. Using Eq. (13), we obtain

$$d^n\sigma = \frac{g^{4n}(2m)^{n-1}}{(2\pi)^{n+1}} \left\{ \prod_{i=1}^{n} \frac{D_i}{f_{ij}} \right\} \frac{\left\{ \prod_{i=1}^{n-1} \Delta(2p\cdot q_i) \right\}}{T_1 T_2 T_3 T_4 ... T_{n-1}} \left\{ \prod_{i=1}^{n} \Delta(2p\cdot q_i) dq_{i0} \right\} \left\{ \prod_{i=1}^{n} \frac{(2p\cdot \tilde{a}_i)^2}{m^2 q_i^4} \frac{d\boldsymbol{q}_{iT} dq_{iz}}{2a_{i0}'} \right\}. \quad (17)$$

The distribution $\Delta(p\cdot q_i)$ can be written as

$$\Delta(2p\cdot q_i) = \frac{1}{p_0 + p_z} \Delta\left(q_{i0} - q_{iz} - \frac{-(p_0 - p_z)(q_{i0} + q_{iz}) + 2\boldsymbol{p}_T \cdot \boldsymbol{q}_T}{p_0 + p_z} \right), \quad (18)$$

which provides the constraint $q_{i0} - q_{iz} \sim 0$. As a consequence, the integral of $|\Delta(2p\cdot q_i)|^2 dq_{i0}$ can be carried out, with the collision time T_i canceling one of the factors in $\Delta(2p\cdot q_i)|_{2p\cdot q_i \to 0}$. The denominators of the boson propagator in Eq. (17) becomes

$$q_i^2 = (q_{i0} + q_{iz})(q_{i0} - q_{iz}) - |\boldsymbol{q}_{iT}|^2 \approx -|\boldsymbol{q}_{iT}|^2. \quad (19)$$

We obtain

$$d^n\sigma = \frac{\alpha^2}{p_z m} \left(\frac{\alpha^2}{p_z^2} \right)^{n-1} \left\{ \prod_{i=1}^{n} \frac{[2mD_i]}{f_{ij}} \right\} \left\{ \prod_{i=1}^{n} \frac{(2p\cdot\tilde{a}_i)^2}{m^2 2a_{i0}'} \frac{dq_{iz} d\boldsymbol{q}_{iT}}{|\boldsymbol{q}_T|^4} \right\}, \quad (20)$$

where $\alpha = g^2/4\pi$, the last two factors are dimensionless, and $d^n\sigma$ has the dimension of $(\alpha^2/p_z^2)^n$.

To find the probability distribution for the longitudinal momentum transfer q_{iz}, we introduce the fractional longitudinal momentum kick

$$x_i = \frac{q_{iz}}{p_z}, \qquad dq_{iz} = p_z dx_i. \quad (21)$$

To investigate the x_i dependence of the factor $(2p\cdot\tilde{a}_i)^2/2a_{i0}'$ in Eq. (20), we note from Eq. (6) that $\tilde{a}_i$ can be written as a function of q_i and a_i,

$$\tilde{a}_i \sim \sqrt{\frac{a_{i0} + m}{a_{i0}' + m}} \frac{q_i}{2} + \frac{a_{i0}' + m + a_{i0} + m}{\sqrt{(a_{i0}' + m)(a_{i0} + m)}} \frac{a_i}{2}. \quad (22)$$

Because of the $\Delta(2p\cdot q_i)$ constraint, the factor $(2p\cdot\tilde{a}_i)^2/2a_{i0}'$ in Eq. (20) becomes

$$\frac{(2p\cdot\tilde{a}_i)^2}{2a_{i0}'} \sim \frac{(a_{i0}' + a_{i0})^2}{2(a_{i0}')^2} \frac{(p\cdot a_i)^2}{a_{i0}} \equiv \kappa_i \frac{(p\cdot a_i)^2}{a_{i0}}. \quad (23)$$

We obtain from Eq. (25)

$$d^n\sigma = \left\{ \frac{\alpha^2}{p_z m} \left(\frac{\alpha^2}{p_z^2} \right)^{n-1} \frac{p_z^n}{m^{2n}} \left(\prod_{i=1}^{n} \frac{[2mD_i]}{f_{ij}} \frac{\kappa_i (p\cdot a_i)^2}{a_{i0}} \right) \right\} \frac{dx_1 dx_2 ... dx_n d\boldsymbol{q}_{1T} d\boldsymbol{q}_{2T} ... d\boldsymbol{q}_{nT}}{|\boldsymbol{q}_{1T}|^4 \, |\boldsymbol{q}_{2T}|^4 \, ... \, |\boldsymbol{q}_{nT}|^4}. \quad (24)$$

The fermion scatterers can possess different initial energies a_{i0} at the moment of their collisions with the energetic jet. In the case when $a_{i0} \ll q_{i0}$, the factor κ_i approaches $1/2 + O(a_{i0}/q_{iz})$ with $(a_{i0}/q_{iz}) \ll 1$. In the other extreme when $a_{i0} \gg q_{i0}$, the factor κ approaches $2 + O(q_{iz}/a_{i0})$ with $(q_{iz}/a_{i0}) \ll 1$. The dependence of κ_i on x_i is weak in either limits and can be neglected in our approximate estimate. We obtain then approximately

$$d^n\sigma \sim \left\{ \text{constant factor} \right\} \frac{dx_1 dx_2 ... dx_n \;\; d\boldsymbol{q}_{1T} d\boldsymbol{q}_{2T} ... d\boldsymbol{q}_{nT}}{|\boldsymbol{q}_{1T}|^4 \, |\boldsymbol{q}_{2T}|^4 \, ... \, |\boldsymbol{q}_{nT}|^4}, \quad (25)$$

where the constant factor is approximately independent of x_i and $\boldsymbol{q}_{iT}$. By symmetry, the fraction of momentum transfers x_i for different scatterers should be approximately the same on the average, and $x_i^{\max}\sim 1/n$. Then as far as x_i is concerned, the average distribution is

$$\left.\frac{dP}{dx_i}\right|_{x_j=x_i,j=1,\dots,n} \sim n\ \Theta\left(\frac{1}{n}-x_i\right), \tag{26}$$

and the average longitudinal momentum fraction is

$$\langle x_i\rangle \sim \frac{1}{2n} \quad \text{or} \quad \langle q_{iz}\rangle \sim \frac{p_z}{2n}. \tag{27}$$

The above results in Eqs. (25)-(27) indicate that in the passage of an energetic fermion making coherent collisions with medium partons, the reaction has a high probability for the occurrence of small values of $|\boldsymbol{q}_{iT}|$. The singularities at $|\boldsymbol{q}_{iT}|\sim 0$ in Eq. (25) correspond to the case of infrared instabilities that may be renormalized, and a momentum cut-off $\Lambda_{\rm cut}$ may be introduced. The scatterers acquire an average longitudinal momentum $\langle q_{iz}\rangle \sim p_z/2n$ that is expected to be much greater than $\langle|\boldsymbol{q}_{iT}|\rangle$. Thus, there is a collective quantum many-body effect arising from Bose-Einstein interference such that the fermion scatterers emerge in the direction of the incident particle, each carrying a fraction of the forward longitudinal momentum of the incident particle that is inversely proportional to twice the number of scatterers, $\langle q_{iz}\rangle \sim p_z/2n$.

4. Bose-Einstein Interference for Coherent Collisions in Non-Abelian Theory

The above considerations for the Abelian theory can be extended to the non-Abelian theory. As an example, we consider a quark jet p making coherent collisions with quarks a_1 and a_2 in the reaction $p + a_1 + a_2 \to p' + a_1' + a_2'$, in the non-Abelian theory. We shall neglect four-particle vertices and loops, which are of higher-orders. The Feynman diagrams are then the same as those in Fig. 1. One associates each quark vertex with a color matrix $T^{(p,1,2)}_{\alpha,\beta}$ where the superscript $p, 1$, or 2 identifies the quark p, a_1, or a_2, and the subscripts α or β give the $SU(3)$ color matrix index. The Feynman amplitude M_a for diagram 1(a) is

$$M_a = -g^4 \bar u(\boldsymbol{p}')T^{(p)}_\beta \gamma_\nu \frac{1}{\not p - \not q_1 - m + i\epsilon'} T^{(p)}_\alpha \gamma_\mu u(\boldsymbol{p}) \frac{1}{q_2^2} \bar u(\boldsymbol{a}_2') T^{(2)}_\beta \gamma_\nu u(\boldsymbol{a}_2) \frac{1}{q_1^2} \bar u(\boldsymbol{a}_1{}') T^{(1)}_\alpha \gamma_\mu u(\boldsymbol{a}_1).$$

The Feynman amplitude M_b for diagram 1(b) is

$$M_b = -g^4 \bar u(\boldsymbol{p}')T^{(p)}_\alpha \gamma_\mu \frac{1}{\not p - \not q_2 - m + i\epsilon'} T^{(p)}_\beta \gamma_\nu u(\boldsymbol{p}) \frac{1}{q_2^2} \bar u(\boldsymbol{a}_2') T^{(2)}_\beta \gamma_\nu u(\boldsymbol{a}_2) \frac{1}{q_1^2} \bar u(\boldsymbol{a}_1{}') T^{(1)}_\alpha \gamma_\mu u(\boldsymbol{a}_1).$$

In the high-energy limit for coherent collisions, the sum of the Feynman amplitudes is

$$M \sim \frac{g^4}{2m}\frac{2p\cdot\tilde a_1}{m}\frac{2p\cdot\tilde a_2}{m}\frac{1}{q_2^2 q_1^2}\left(\frac{T^{(p)}_\beta T^{(p)}_\alpha T^{(2)}_\beta T^{(1)}_\alpha}{2p\cdot q_1 - i\epsilon} + \frac{T^{(p)}_\alpha T^{(p)}_\beta T^{(2)}_\beta T^{(1)}_\alpha}{2p\cdot q_2 - i\epsilon}\right). \tag{28}$$

We can rewrite the product of the color matrices for the quark jet p as

$$T^{(p)}_\beta T^{(p)}_\alpha = \frac{1}{2}\left([T^{(p)}_\beta, T^{(p)}_\alpha]_+ + [T^{(p)}_\beta, T^{(p)}_\alpha]_-\right), \tag{29}$$

$$T^{(p)}_\alpha T^{(p)}_\beta = \frac{1}{2}\left([T^{(p)}_\beta, T^{(p)}_\alpha]_+ - [T^{(p)}_\beta, T^{(p)}_\alpha]_-\right). \tag{30}$$

For coherent collisions, the Feynman amplitude is then

$$M \sim \frac{g^4}{2m}\frac{2p\cdot\tilde{a}_1}{m}\frac{2p\cdot\tilde{a}_2}{m}\frac{1}{q_2^2q_1^2}\left\{(\mathcal{M}_1+\mathcal{M}_2)\frac{[T_\beta^{(p)},T_\alpha^{(p)}]_+T_\beta^{(2)}T_\alpha^{(1)}}{2}+(\mathcal{M}_1-\mathcal{M}_2)\frac{[T_\beta^{(p)},T_\alpha^{(p)}]_-T_\beta^{(2)}T_\alpha^{(1)}}{2}\right\}, \quad (31)$$

$$\mathcal{M}_1+\mathcal{M}_2 = i\Delta(2p\cdot q_1)+i\Delta(2p\cdot q_2), \quad \text{and} \quad \mathcal{M}_1-\mathcal{M}_2=\frac{4p\cdot q_1}{(2p\cdot q_1)^2+\epsilon^2}. \quad (32)$$

¿From the above analysis, we find that the color degrees of freedom in QCD bring in additional properties to the Feynman amplitudes. Bose-Einstein symmetry with respect to the interchange of gluons in QCD involves not only the space-time exchange symmetry but also color index exchange symmetry. The total exchange symmetry can be attained with symmetric space-time amplitudes and symmetric color index factors as in the $(\mathcal{M}_1+\mathcal{M}_2)$ term in Eq. (31). The total symmetry can also be attained with space-time antisymmetry and color index antisymmetry, as in the $(\mathcal{M}_1-\mathcal{M}_2)$ term in Eq. (31) .

For the space-time symmetric and color index exchange symmetric component, the Feynman amplitude is equal to the Abelian Feynman amplitude multiplied by a color factor. It will exhibit the same degree of Bose-Einstein interference as in the Abelian theory. Previous analysis on the longitudinal momentum transfer of recoiling fermions in the Abelian theory can be applied to the non-Abelian theory for this space-time symmetric and color index exchange symmetric component. There is thus a finite probability for the presence of constraints to lead to recoiling quarks receiving significant momentum kicks along the direction of the incident quark jet.

The above considerations for jets and scatterers can be extended to cases involving gluon jets and gluon scatterers. The results are similar but with a small modification on the distribution for gluon scatterers [1].

5. Signatures for Bose-Einstein Interference in the Passage of a Jet

From the results in the above sections, the occurrence of the Bose-Einstein interference in the coherent collisions of a jet with medium partons possesses the following characteristics:

(1) The Bose-Einstein interference is a quantum many-body effect. It occurs only in the coherent multiple collisions of the fast jet with two or more scatterers, $n \geq 2$.

(2) Each scatterer has a transverse momentum distribution of the type $1/|\boldsymbol{q}_T|^4$, which peaks at small values of $|\boldsymbol{q}_T|$.

(3) Each scatterer acquires a longitudinal momentum kick q_z along the incident jet direction that is approximately inversely proportional to twice the number of scatterers, with $\langle q_z\rangle \sim p_z/2n$.

(4) As a consequence, the final effect is the occurrence of collective recoils of the scatterers along the jet direction.

In a high energy central collision of heavy nuclei, both jets and a dense medium are produced soon after the collision and jets propagate in the dense medium. The mean free path for the jet in the medium may be small compared with the longitudinal coherent length to provide the appropriate environment for coherent collisions. It is of interest to search for evidence for Bose-Einstein interference in the passages of the jet in the dense medium. In such a search, we need to separate out jet fragments from recoiling scatterers among the detected hadrons. Such a separation is indeed kinematically possible in the $\Delta\phi$-$\Delta\eta$ correlation measurements where $\Delta\phi = \phi(\text{associated}) - \phi(\text{trigger})$ is the azimuthal angle difference between the produced hadron pair, and $\Delta\eta = \eta(\text{associated}) - \eta(\text{trigger})$ is the pseudorapidity difference. We show in Fig. 2 the $\Delta\phi$-$\Delta\eta$ correlation data from the STAR Collaboration [23] for the most central AuAu collision at RHIC with a high p_T trigger [23]. Jet fragments are distributed in a small cone of $(\Delta\phi, \Delta\eta)\sim 0$ [22]-[25]. Medium particles are distributed in the "ridge" part at $\Delta\phi \sim 0$ (Fig. 2) along the

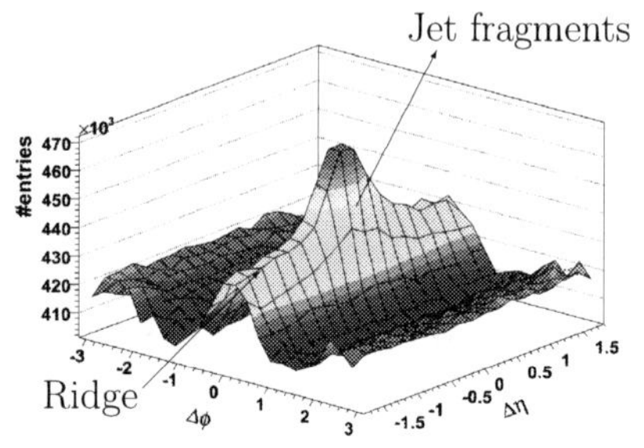

Figure 2. $\Delta\phi$-$\Delta\eta$ correlation of produced hadrons in the most central AuAu collision at RHIC with a high p_T trigger, measured by the STAR Collaboration [23].

$\Delta\eta$ axis. In particular, ridge particle with $|\Delta\eta| > 0.6$ and $\Delta\phi \sim 0$ along the ridge (Fig. 2) can be identified as medium scatterer partons for the following reasons:

(i) The yield of the ridge particles increases approximately linearly with the number of participants [23].
(ii) The yield of the ridge particles is nearly independent of (i) the flavor content, (ii) the meson/hyperon character, and (iii) the transverse momentum p_T (above 4 GeV) of the jet trigger [23, 24].
(iii) The ridge particles have a temperature (inverse slope) that is similar (but slightly higher) than that of the inclusive bulk particles, but lower than the temperature of the near-side jet fragments [23]
(iv) The baryon/meson ratio of the ridge particles is similar to those of the bulk hadrons and is quite different from those in the jet fragments [25].
(v) The Wigner function analysis indicates that the momentum distribution of medium partons has a rapidity plateau structure even at the early stage of flux tube fragmentation [11].

With the medium scatterers (ridge particles) separated from the incident high-p_T jet fragments, the occurrence of the Bose-Einstein interference can be characterized by the collective recoils of the scatterers (the ridge particles) along the jet direction. The collective recoils will lead to the $\Delta\phi \sim 0$ correlation of the ridge particles with the high-p_T trigger. As shown in Fig. 2, such collective recoils of the medium particles have indeed been observed in $\Delta\phi$-$\Delta\eta$ correlations of produced hadrons in AuAu collisions at RHIC by the STAR Collaboration [22]-[25]. Similar $\Delta\phi$-$\Delta\eta$ correlations have also been observed by the PHENIX Collaboration [2, 30, 31], and the PHOBOS Collaboration [32]. Quantitatively, the collective recoils of the kicked medium partons have been encoded into the longitudinal momentum kick $\langle q_z \rangle$ of the momentum kick model analysis that yields the observed $\Delta\phi$, $\Delta\eta$, and p_T dependencies of the angular correlations [6]-[13].

It is of interest to examine the relationship $\langle q_z \rangle \sim p_z/2n$ between n and $\langle q_z \rangle$, when such a collective momentum kick occurs. For the most central AuAu collisions at $\sqrt{s_{NN}} = 200$ GeV at RHIC, we previously found from the momentum kick model analysis that $\langle n \rangle \sim 6$ and $q_z \sim 1$ GeV [8]. In another momentum kick model analysis for the highest multiplicity pp collisions at $\sqrt{s_{NN}} = 7$ TeV at the LHC, we previously found that $\langle n \rangle \sim 2.4$ and $q_z \sim 2$ GeV. The experimental data give a longitudinal momentum transfer $\langle q_z \rangle$ that is approximately inverse proportional to the number of scatterers $\langle n \rangle$, in rough agreement with the signatures discussed above.

With regard to the threshold $n \geq 2$ for the occurrence of the Bose-Einstein interference, the presence of a threshold implies a sudden increase of the ridge yield as a function of centrality,

Journal of Physics: Conference Series **387** (2012) 012009 doi:10.1088/1742-6596/387/1/012009

as represented by the number of participants. Although the experimental data with a high p_T trigger appear to be consistent with the presence of thresholds, the large error bars and the scarcity of the number of data points in the threshold regions preclude a definitive conclusion.

However, threshold effects for the ridge yield (2D Gaussian yield) as a function of the number of participants, $N_{\rm part}$, have been observed in another angular correlation measurements with a low-p_T trigger from the STAR Collaboration [26, 27, 28, 29]. We note previously that a fast jet parton possesses low-p_T jet fragments and a minimum-p_T-biased low-p_T trigger can also indicate the passage of a fast parent jet [13]. As a consequence, ridge particles will also be associated with a low-p_T trigger. The sudden increase of the amplitude and the peak η width of the ridge yield as a function of $N_{\rm part}$ may indicate the presence of a threshold for the ridge yield as a function of centrality [1].

6. Conclusions and Discussions

In the collision of a fast particle with many medium particles, the set of multiple collisions can occur incoherently as a sequence of n two-body collisions or coherently as a single $(1+n)$-body collision. From the binary collision data, one obtains the longitudinal correlation length $\Delta z_{\rm coh} \sim 2\hbar p_0/q_T^2$. If the medium is so dilute, the binary cross section so small, and the incident particle so slow that $\lambda \gg \Delta z_{\rm coh}$, then the set of multiple collisions consists of a sequence of incoherent collisions. On the other hand, if the medium is so dense, the binary cross section so large, and the incident particle is so energetic such that $\Delta z_{\rm coh} \gg \lambda$, then the set of multiple collisions consists of a single coherent $(1+n)$-body collision.

In central collisions of heavy nuclei at RHIC energies, the environment of the produced medium favors the occurrence of coherent collisions between the jet and medium particles. The dynamics is governed by Feynman diagrams linking the incident jet with the scatterers as a connected unit. For a set of initial states and a set of final states in such a coherent collision, there are many different trajectories in the sequences of collisions along $\Delta z_{\rm coh}$ at which various virtual bosons are exchanged. By Bose-Einstein symmetry, the total Feynman amplitude is the sum of all Feynman amplitudes with all interchanges of the virtual boson vertices.

Remarkably, in high-energy collisions, the summation of these Feynman amplitudes from different ways of ordering the virtual boson vertices interfere with each other, resulting in the cancellation of some parts of the Feynman amplitudes and leaving only the other parts of sharp distributions. The longitudinal momentum transfer to each scatterer is then constrained to be the same as the energy transfer. There is a substantial flow of the longitudinal momentum from the jet to the scatterers. Furthermore, the transverse momentum distribution favors small values of q_T. As a consequence, the scatterers recoil collectively along the jet direction, leading to a correlation of the scatterers and the jet around $\Delta\phi \equiv \phi(\text{scatterer}) - \phi(\text{trigger jet}) \sim 0$.

For the coherent collision of an energetic parton with parton scatterers in non-Abelian cases, we find that the complete Bose-Einstein symmetry in the exchange of virtual gluons consists not only of space-time exchange symmetry but also color index exchange symmetry. There is always a space-time symmetric and color-index symmetric component of the Feynman amplitude that behaves in the same way as the Feynman amplitude in the Abelian case, and the corresponding recoiling partons behave in the same way as in the Abelian case. There is thus a finite probability for the parton scatterers to emerge collectively along the incident trigger jet direction, each carrying a significant fraction of the longitudinal momentum of the incident jet.

The collective recoils will lead to the $\Delta\phi \sim 0$ correlation of the ridge particles with the high-p_T trigger. Such a signature of the Bose-Einstein interference may have been observed in the $\Delta\phi \sim 0$ correlation in the angular correlation measurements of produced hadron pairs in central AuAu collisions at RHIC [22]-[31]. The centrality dependence of the ridge yield in AuAu collisions at $\sqrt{s_{NN}} = 200$ GeV with a low-p_T trigger [26, 27, 28, 29] may also be consistent with the presence of a ridge threshold at $n = 2$, as expected in the quantum many-body effect of

Bose-Einstein interference.

The collective recoils of the scatterers from the Bose-Einstein interference may be the origin of the the longitudinal momentum kick along the jet direction postulated in the momentum kick model that has been quite successful in the analysis of the angular correlations of hadrons produced in high-energy heavy-ion collisions [6]-[13]. It is interesting to note that the present Bose Einstein interference in coherent collisions is consistent with previous results on the Bose-Einstein interference in the emission of real photons and gluons in high-energy interactions and in the sum of the ladder and cross-ladder loop diagrams in the collision of two particles [15]-[21].

Acknowledgment

The author would like to thank Profs. Vince Cianciolo, Horace W. Crater, C. S. Lam, and Jin-Hee Yoon for helpful discussions. This research was supported in part by the Division of Nuclear Physics, U.S. Department of Energy.

References

[1] Wong C Y 2011 *arXiv:1112.1091*
[2] Adare A *et al.* (PHENIX Collaboration) 2008 *Phys. Rev. C* **78**, 014901
[3] Bjorken J D 1982 *Fermilab-Pub-82-059-THY*
[4] Gyulassy M and Wang X N 1994 *Nucl Phys* **B420** 583
[5] Adler S S *et al.* (PHENIX Collaboration) 2004 *Phys Rev C* **69** 034910; Adam J *et al.* (STAR Collaboration) 2003 *Phys Rev Lett* **91** 172302; Back B B *et al.* (PHOBOS Collaboration) 2004 *Phys Lett* **B578** 297; Arsene I *et al.* (BRAHMS Collaboration) 2003 *Phys Rev Lett* **91** 072305; Abelev B I *et al.* (STAR Collaboration) 2007 *Phys Rev Lett* **98** 192301; Adare S S *et al.* (PHENIX Collaboration) 2006 *Phys Rev Lett* **97** 252002
[6] Wong C Y 2007 *Phys Rev C* **76** 054908
[7] Wong C Y 2008 *Chin Phys Lett* **25** 3936
[8] Wong C Y 2008 *J Phys G* **35** 104085
[9] Wong C Y 2008 *Phys Rev C* **78** 064905
[10] Wong C Y 2009 *Phys Rev C* **80** 034908
[11] Wong C Y 2009 *Phys Rev C* **80** 054917
[12] Wong C Y 2009 *Nonlin Phenom Complex Syst* **12** 315; arXiv:0911.3583
[13] Wong C Y 2011 *Phys Rev C* **84** 024901
[14] Cheng H and Wu T T 1987 *Expanding Protons: Scattering at High Energies* (MIT Press)
[15] Cheng H and Wu T T 1969 *Phys Rev* **186** 1611
[16] Feng Y J, Hamidi-Ravari O, and Lam C S 1996 *Phys Rev D* **54** 3114
[17] Feng Y J, Hamidi-Ravari O, and Lam C S 1997 *Phys Rev D* **55** 4016
[18] Lam C S 1996 *Lectures given at the First Asia Pacific Workshop on Strong Interactions* Taipei August 1st to 27th; hep-ph/9704240
[19] Lam C S and Liu K F 1997 *Nucl Phys* **B 483** 514
[20] Lam C S and Liu K F 1997 *Phy Rev Lett* **79** 597
[21] Lam C S 1997 *Chin Jour Phys* **35** 758; hep-ph/9805210
[22] Adams J *et al.* for the STAR Collaboration 2005 *Phys Rev Lett* **95** 152301
[23] J. Putschke (STAR Collaboration) 2007 *J Phys G* **34** S679
[24] Bielcikova J (STAR Collaboration) 2007 *J Phys G* **34** S929
[25] van Leeuwen M (STAR Collaboration) 2009 *Eur Phys J C* **61** 569
[26] Daugherity M (STAR Collaboration) 2008 *J Phys G* **35** 104090
[27] Ray L 2008 *Talk presented at Tamura Symposium on Heavy Ion Physics* The University of Texas at Austin; http://www.ph.utexas.edu/~molly/tamura/
[28] Kettler D (STAR Collaboration) 2009 *Eur Phys J C* **62** 175
[29] Agakishiev G *et al.* (STAR Collaboration) 2011 *arxiv:1109.4380*
[30] McCumber M P (PHENIX Collaboration) 2008 *J Phys G* **35** 104081
[31] Jia J (PHENIX Collaboration) 2008 *J Phys G* **35** 104033
[32] Wenger E (PHOBOS Collaboration) 2008 *J Phys G* **35** 104080

XXXV Symposium on Nuclear Physics IOP Publishing
Journal of Physics: Conference Series **387** (2012) 012010 doi:10.1088/1742-6596/387/1/012010

Conventional vs gauge couplings in elastic πN scattering

A. Mariano, C. Barbero, D. Badagnani

Departamento de Física, Facultad de Ciencias Exactas, Universidad Nacional de La Plata, Argentina.
Instituto de Física La Plata, CONICET, Argentina.

E-mail: `mariano@fisica.unlp.edu.ar`

Abstract. We perform a review on the importance of the Δ(1232 MeV) resonance in different processes, analyze the existing problems with the fields used to describe it theoretically and finally we make an analysis on the use of $\pi N\Delta$ alternative couplings in πN scattering. We compare numerically results for the elastic cross section obtained by using the conventional couplings, already adopted in several reaction calculations, with those obtained with the so called "spin-3/2 gauge invariant" vertexes suggested recently. Confronting with experimental cross section data, we see that these results are by no means equivalent and that the differences between them cannot be compensated by a readjustment of free parameters of background contributions. We find that the conventional couplings work better than the gauge ones in describing the data.

1. Introduction

We can ask ourselves, why to put attention on the isobar Δ(1232 MeV) resonance? The immediate answer would be: since it is the first excited state of the nucleon(N) and has a prominent role in strong interactions. It dominates the pion-production phenomena being excited in $\pi N \to \pi' N'$, $\pi N \to \pi' N' \gamma$[11] scattering, which are useful reactions in order to check different isobar models regards Lorentz invariance, unitarity, and gauge invariance. It is also present in the $\gamma N \to \pi N$ photoproduction process[2], important to determine the deformation in the 3 quark(q) constitutive wave function. Also, it appears in $eN \to e' N' \pi$ electroproduction scattering [3] that is a reaction important to get vector form factors and participates in the $\nu N \to l N' \pi$ weak production [4], this being the main experimental background in neutrino oscillation experiments. On the other side, the Δ resonance dominates many nuclear phenomena at energies above the π-production threshold. In cosmology, it is largely responsible for the "GZK cut-off" [5], drooping of cosmic ray rate due excitation of Δ by scattering of CMB photons. High precision measurements of the $N \to \Delta$ transition was possible at LEGS, BATES, ELSA, MAMI and JLAB (several experiments on scattering by electromagnetic probes). The experimental effort, has been accompanied by exciting developments in dynamical models, lattice QCD and chiral effective-field theories.

The Δ(1232) was discovered 60 years ago by Fermi and Colaborators at the Chicago cyclotron (now Fermilab), being it mass, width, spin and isospin, $m = 1232$ MeV$\simeq m_N$+300 MeV, $\Gamma = 120$

Journal of Physics: Conference Series **387** (2012) 012010 doi:10.1088/1742-6596/387/1/012010

MeV ($\tau = 10^{-23}$ sec), $S = T = 3/2$, respectively. Within the SU(6) quark model the N appears as the ground state of a 3q system in a confining potential. In the same way the Δ excitation in πN scattering, for example $\pi(u\overline{d})p(uud) \to \Delta^{++}(uuu) \to \pi'(u\overline{d})p'(uud)$, can be seen as a $d\overline{d}$ annihilation. It then decays 99% of time in πN channel and 1% in the $\gamma\pi$ electromagnetic one, being this excitation essentially a spin flip magnetic dipole one (see Figure 1). Nevertheless, if we considered D admixtures in the N or Δ wave functions, a quadrupole transition is present.

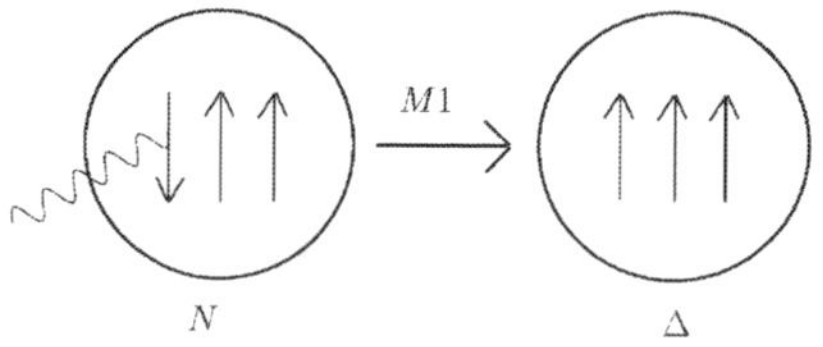

Figure 1. $M1$ Excitation of the Δ.

From the theoretical point of view, we know that pions are the Goldstone bosons of the spontaneously broken chiral symmetry of QCD and as the coupling with them goes as p_π, at low energies perturbative expansion in powers of it is possible, we having the Chiral Perturbation Theory [7]. On the other hand the resonance parameters can be extracted from reactions by using unitary isobar models: unitarized tree-level calculations based on phenomenological Lagrangians. Nevertheless "pion cloud" effects, which represents the coupling of a pion to quarks, could only be comprehensively studied within dynamical models based in T-matrix calculations.

2. The Rarita-Schwinger field

For implementing any of the mentioned approaches we need the kinetic $\mathcal{L}_\Delta$ and interacting $\mathcal{L}_{\pi N\Delta}$ Lagrangians. The Rarita Schwinger (RS) spinor ψ_μ, which represents mathematically the Δ, is taken as an element of the non-unitary representation $[(1/2,0) \oplus (0,1/2)] \otimes [(1/2,1/2) = (1/2,0) \otimes (0,1/2)]$ of the Lorentz group. That is, $\psi_\mu \equiv \psi \otimes W_\mu$, being ψ a Dirac spinor and W_μ is a 4-vector with spin $S = 0,1$ in the rest or helicity frames. Omitting for a moment the Lorentz structure and putting attention only in the angular momentum we have representations $D(1/2) \otimes D(0) = D(1/2)$ and $D(1/2) \otimes D(1) = D(3/2) \oplus D(1/2)$, what enclose $2(S = 1/2) + 4(S = 3/2) + 2(S = 1/2) + antiparticles = 16$ degrees of freedom (dof). $\psi_\mu(x)$ satisfies the Dirac and Klein Gordon equations and additionally those states built with $W_\mu(S = 1)$, satisfy the Proca one. From the 16 states only 8 satisfy the subsidiary conditions $\partial^\mu \psi_\mu = \gamma^\mu \psi_\mu = 0$, what are seen as making zero "projections" of ψ_μ on spin-1/2 subspaces to get right dof counting for the free case.

The RS field is a "constrained" dynamical system supplemented by constraints or subsidiary conditions , being the Lagrangian from which we can get the equation of motion plus the constraints

$$\mathcal{L}_{free} = \overline{\psi}_\mu(x) \left\{ i\partial_\alpha \Gamma^\alpha_{\mu\nu} - mB_{\mu\nu} \right\} \psi^\nu(x), \tag{1}$$

where

$$\begin{aligned} \Gamma^\alpha_{\nu\mu} &= g_{\mu\nu}\gamma^\alpha + \frac{1}{3}\gamma_\mu\gamma^\alpha\gamma_\nu - \frac{1}{3}(\gamma_\mu g^\alpha_\nu + g^\alpha_\mu \gamma_\nu), \\ B_{\nu\mu} &= g_{\mu\nu} - \frac{1}{3}\gamma_\mu\gamma_\nu , \end{aligned} \tag{2}$$

proposed by Rarita and Schwinger in 1941. $\mathcal{L}_{free}$ includes the constraints and Γ, B do not mix 3/2 with 1/2 states, then it only fix 3/2 component of ψ_μ. Then, the contact transformation $\psi^\mu \to \psi'^\mu = R(a)^{\mu\nu}\psi_\nu \equiv (g^{\mu\nu} + a\gamma^\mu\gamma^\nu)\psi_\nu$, only affects $\psi_{1/2\mu}$ components of ψ_μ and let $\mathcal{L}_{free}$ invariant we having a whole family of valid one parameter Lagrangians ($a = 1/2(1+3A)$)

$$\mathcal{L}_{free}(A) \;=\; \overline{\psi}(x)R(A)\left\{i\partial_\alpha\Gamma^\alpha - mB\right\}R(A)\psi(x), \tag{3}$$

where for $A = -1/3$ we recuperate the RS one and where some indexes are omitted by simplicity. Finally the spin-3/2 propagator $G(p, A)$ can be obtained as usually giving

$$\begin{aligned} G(p,A) &= R^{-1}(A)G\left(p,-\frac{1}{3}\right)R^{-1}(A), \\ G\left(p,-\frac{1}{3}\right)_{\mu\nu} &= -\left[\frac{\not p + m}{p^2-m^2}\hat{P}^{3/2}_{\mu\nu} + \frac{2}{m^2}(\not p + m)(\hat{P}^{1/2}_{11})_{\mu\nu} + \frac{\sqrt{3}}{m}(\hat{P}^{1/2}_{12} + \hat{P}^{1/2}_{21})_{\mu\nu}\right], \end{aligned} \tag{4}$$

this last being the RS propagator and P^k_{ij} the projectors on the $k = 3/2, 1/2$ sector of the representation space and with $i,j = 1,2$ indicating sub-sectors of the 1/2 subspace [8]. At difference of the on-shell case where the subsidiary conditions select only the 3/2 states, when Δ is off-shell ($p^2 \neq m^2$) the 1/2 ones appear. One can wonder if this situation is exclusive of the RS field. And the answer is no, for example the W propagator appearing in $\pi \to W \to \overline{\nu}\mu$ or pion-pole terms in $\nu N \to \mu N'\pi$ reactions (see Figure 2)

$$\Delta^{\mu\nu}(p) = -\left[\frac{P_1^{\mu\nu}(p)}{p^2-m^2} + \frac{P_0^{\mu\nu}(p)}{m^2}\right], \tag{5}$$

has P_0 and P_1 projectors on the $S = 0$ and $S = 1$ sectors, respectively. Again here we have an off-shell lower spin contribution as for RS case.

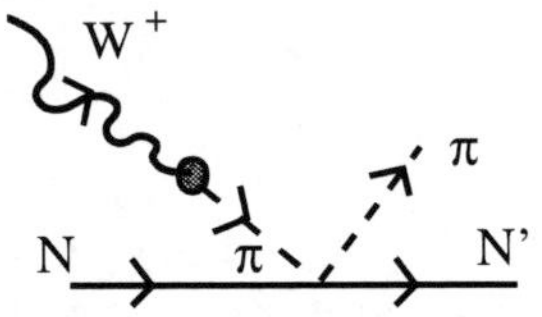

Figure 2. W meson propagation between the lepton weak vertex and a conversion to a pion in a pion production reaction.

3. Interactions

$\mathcal{L}_{int}$ must be invariant under contact transformations and lead to A-independent amplitudes. We analyze the strong $\pi(\phi)N(\Psi)\Delta(\psi_\mu)$ vertex, as appears in the amplitude of Figure 3, but the analysis can be extended to another type of interactions. We will analyze two different models for the $\pi N\Delta$ vertex, the "conventional" (C) and the "3/2-gauge invariant" (G) one. The C vertex is inspired in the nonlinear realization of the chiral symmetry[9], which leads to a derivative of the pion field

$$\mathcal{L}_{int_C} = \frac{f_{\pi N\Delta}}{m_\pi}\bar{\Psi}\partial^\mu\phi R\left(\frac{1}{2}(1+3A)\right)_{\mu\nu}\psi^\nu + \text{h.c.}, \tag{6}$$

and as this vertex couples to the spin-1/2 off-shell Δ sector and one must pursue the constraints analysis to generate new subsidiary conditions, this fact is viewed as a dof counting inconsistence.

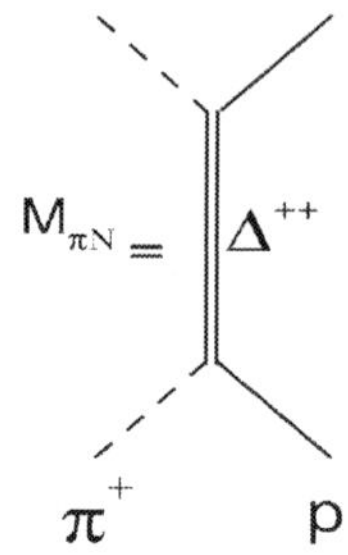

Figure 3. Δ-pole contribution to π^+p elastic scattering.

Nevertheless, the $W \leftrightarrow \pi$ vertex also couples to the spin-0 sector, more it is impossible for a pion to decay without coupling to the off-shell spin-0 piece of the W propagator. The G vertex is inspired by the fact that in the massless case $\mathcal{L}_{free}$ is invariant under the spin-3/2 gauge transformation[10] $\psi_\mu(x) \to \psi_\mu(x) + R^{-1}(\frac{1}{2}(1+3A))_{\mu\nu}R(-1)^{\nu\mu'}\partial_{\mu'}\chi(x)$, and one looks for a Lagrangian respecting the same symmetry[1]

$$\mathcal{L}_{int_G}(A) = \frac{f_{\pi N\Delta}}{m_\pi m}\bar{\Psi}\partial_\mu\phi^\dagger\epsilon^{\mu\nu\alpha\beta}\gamma_\beta\gamma_5 R(-\frac{1}{2}(A+1))_{\nu\sigma}\partial_\alpha\psi^\sigma + \text{h.c.}. \tag{7}$$

This vertex contracted with the Δ momentum is zero, which in turn nullifies the contraction with $P^{\frac{1}{2}\mu\nu}_{ij}$ avoiding the coupling with the spin-1/2 sector in (4)(this is easy to check for $A=-1$). Nevertheless, when we couple a photon making the minimal substitution $\partial_\mu \to \partial_\mu - iq_\Delta A_\mu$, the obtained electromagnetic Lagrangians are no more spin-3/2 gauge invariant appearing terms $q_\Delta\partial_\mu\overline{\chi}\epsilon^{\mu\nu\alpha\beta}A_\alpha\gamma_\beta\gamma_5\psi_\nu + ...$, this indicating that both gauge symmetries cannot coexist.

4. π^+p cross section calculation

Viewing the evident complexity of the problem and since the C vertex constraints problem is not present in perturbative calculations, leaving for a moment the G vertex 3/2-electromagnetic gauge problem, we simply compare C and G couplings in the case of πN elastic scattering within an isobar model. The different contributions considered to built the amplitude are depicted in Figure 4. We put attention in the Δ-pole amplitude contribution (the Δ-cross one can be

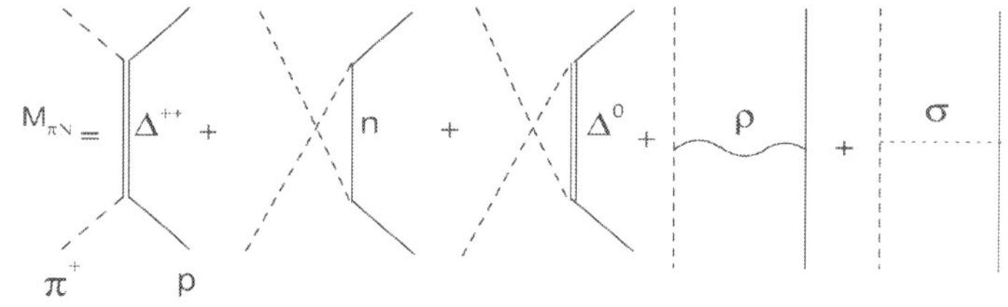

Figure 4. Feynman graphs corresponding to different contributions to the elastic π^+p scattering amplitude.

obtained on the same footing) since the other ones can be constructed as usually from the

[1] Now the corresponding propagator should be $G^{\mu\nu}(p,A) = R^{-1}\left(-\frac{1}{2}(A+1)\right)^\mu_{\ \alpha} G^{\alpha\beta}(p,-1)R^{-1}\left(-\frac{1}{2}(A+1)\right)^\nu_{\ \beta}$.

Journal of Physics: Conference Series **387** (2012) 012010 doi:10.1088/1742-6596/387/1/012010

corresponding hadronic Lagrangians [11], we have

$$\begin{aligned}
\mathcal{M}^{C}_{\Delta^{++}} &= \frac{f^2_{\pi N\Delta}}{m^2_\pi}\overline{u}(p'_p, m'_s)p^\mu_\pi G_{\mu\nu}(p, -\frac{1}{3})p^\nu_\pi u(p_p, m_s), \\
\mathcal{M}^{G}_{\Delta^{++}} &= \frac{f^2_{\pi N\Delta}p^2}{m^2_\pi m^2}\overline{u}(p'_p, m'_s)p^\mu_\pi(-)\frac{\not p + m}{p^2 - m^2}\hat{P}^{\frac{3}{2}}_{\mu\nu}p^\nu_\pi u(p_p, m_s).
\end{aligned} \tag{8}$$

To overcome the singular behavior of the resonant amplitude at $p^2 = m^2$ we use the complex mass scheme (see Ref.[11]) making the replacement $m^2 \to m^2 - im\Gamma$. The Δ mass, width and coupling constants $g_\sigma = g_{\sigma\pi\pi}g_{\sigma NN}$ and $f_{\Delta N\pi}$ are left as the only free parameters to be determined from the fitting of the total cross section of π^+p scattering to the data [12]. The

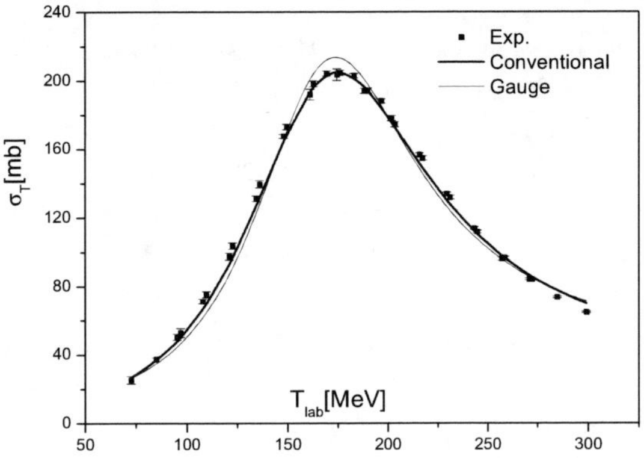

Figure 5. Elastic π^+p total cross section calculated with the amplitude with $\mathcal{M}^{C}_{\Delta^{++}}$ and $\mathcal{M}^{G}_{\Delta^{++}}$.

mass of the hypothetical σ meson that runs on the range $400 - 1000$ MeV region [13] is strongly correlated with the g_σ value (see Ref.[11]). We have moved m_σ on the full mentioned range and have looked for the value of g_σ to get the best fitting for each adopted m_σ, comparing at the same time fittings done with different m_σ. We find that for $m_\sigma = 450$ MeV or $m_\sigma = 650$ MeV, depending on the model for $\mathcal{M}_{\Delta^{++}}$, we get in each case the best fitting and we are sure that for these values we have the better description with each coupling. For the C couplings the best fitting corresponds to $m_\sigma = 650$ MeV, and we get : $f^2_{\Delta N\pi}/4\pi = 0.317 \pm 0.003$, $m_\Delta = 1211.2 \pm 0.4$ MeV, $\Gamma = 88.2 \pm 0.4$ MeV, $g_\sigma/4\pi = 1.50 \pm 0.12$, and $\chi^2/dof = 4.5$, while for the G coupling we use $m_\sigma = 450$ MeV, and get 0.278 ± 0.002, 1211.6 ± 0.3 MeV, 76.62 ± 0.25 MeV, 1.00 ± 0.05 and 13.5, respectively. Results are shown in Figure 5. With these obtained parameters we also show the predicted differential cross section at two fixed T_{lab} energy values. Our results are compared with available data for both C and G couplings in the Figure 6. Note that following the philosophy of effective Lagrangian models we have not introduced *ad-hoc* form factors for the vertices, since the description of low-energy hadron interactions must incorporate only the (structureless) relevant degrees of freedom. In addition, our model does not consider the final state rescattering between the pion and the nucleon. These last two effects surely improve the fitting, but more free parameters should be considered increasing the model-dependency of the calculation and obscuring the C-G comparison.

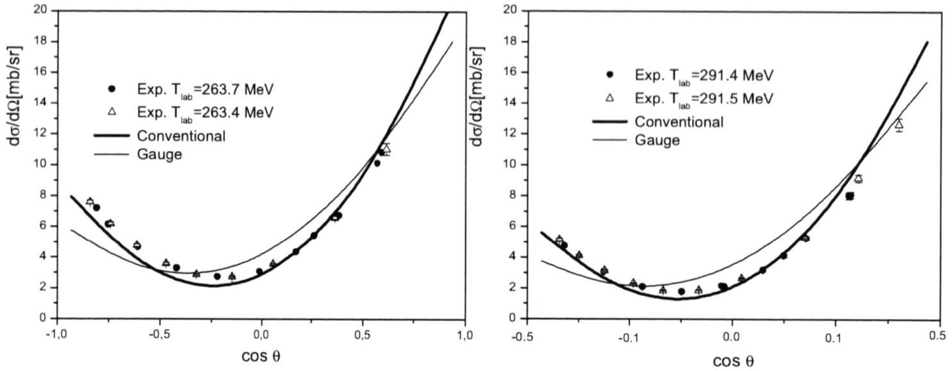

Figure 6. Differential $\pi^+ p$ cross section calculated with the conventional (C) and gauge (G) amplitudes for $T_{lab} = 263.7, 291.4$ MeV. Circles and triangles indicate experimental data from Refs. [14] and [15], respectively.

5. Conclusions

The so called "inconsistent" off-shell propagation of the 1/2 lower spin components of the Δ field is clearly present in other cases, as it is for example the W boson induced reactions. Within a simple isobar model including resonant and background terms, the fitting achieved with the C couplings are clearly better than those obtained with G ones. Seems not possible accommodate the parameters of the σ meson (those of ρ are fixed in both approaches by low energy phenomenology) to get identical results with both types of couplings. Giving the problems with the constraint analysis within the C couplings and those of coexistence between the spin-3/2 and electromagnetic gauge invariance within the G ones, one can appreciate the complexity of the problem and that is not closed.

References

[1] López Castro G and Mariano A 2001 *Nuc. Phys. A* **697** 440.
[2] Mariano A 2007 *Jour. Phys. G***34** 1627.
[3] Sato T and Lee T -S H 2001 *Phys. Rev. C* **63** 055201.
[4] Barbero C, López Castro G and Mariano A 2008 *Phys. Lett. B***664** 70.
Mariano A, Barbero C and López Castro G 2011 *Nuc. Phys. A***849** 218.
[5] Mucke A, Rachen J P, Engel R, Protheroe R J, Stanev T 1999 *Publ. Astron. Soc. Austral.* **16** 160.
[6] Anderson H L, Fermi E, Long E A, Nagle D E 1952 *Phys. Rev.* **85** 936.
[7] Gasser J, Leutwyler H 1984 *Ann. Phys.* **158** 142.
[8] de Jong F and Malfliet R 1992 *Phys. Rev. C* **46** 2567.
[9] L. M. Nath, B. Etemadi and J. D. Kimmel, Phys Rev **D3** (1971) 2153.
[10] V. Pascalutsa, Phys. Lett. **B503** (2001) 85.
[11] G. López Castro and A. Mariano, Nucl. Phys. **A697** (2001) 440.
[12] Pedroni E, it et al 1978 *Nucl. Phys. A* **300** 321.
[13] Nakamura K, *et al* 2010 *Particle Data Group, Jour. Phys. G* **37** 1.
[14] Bussey P J, et al 1973 *Nuc. Phys. B* **58** 363.
[15] Sadler M E, et al 1987 *Phys. Rev. D***35** 2718.

Journal of Physics: Conference Series **387** (2012) 012011 doi:10.1088/1742-6596/387/1/012011

Strangeness of the proton

Roelof Bijker

Instituto de Ciencias Nucleares, Universidad Nacional Autónoma de México, A.P. 70-543, 04510 México, D.F., México

E-mail: bijker@nucleares.unam.mx

Jacopo Ferretti

I.N.F.N., Sezione di Genova, via Dodecaneso 33, I-16146, Italy

E-mail: jacopo.ferretti@ge.infn.it

Elena Santopinto

I.N.F.N., Sezione di Genova, via Dodecaneso 33, I-16146, Italy

E-mail: santopinto@ge.infn.it

Abstract. The contribution of strange quarks to the radius and the magnetic moment of the proton is discussed in the framework of the unquenched quark model in which the effects of the $s\bar{s}$ pairs (strange sea) are taken into account through a $^3P_0$ quark-antiquark pair creation mechanism. The results for the the strange magnetic moment and the strangeness radius of the proton are found to be compatible with zero, in agreement with the latest experimental results from parity-violating electron scattering experiments and recent lattice calculations.

1. Introduction

Electromagnetic and weak form factors are key ingredients to the understanding of the internal structure of the nucleon, since they contain the information about the distributions of electric charge and magnetization. Deep-inelastic scattering experiments have demonstrated that the structure of the proton cannot be described by its uud valence structure alone: the valence quarks carry about half of the proton momentum and about one third of the proton spin, and the observed violation of the Gottfried sum rules indicates that there are more $\bar{d}$ quarks in the proton than there are $\bar{u}$. The contribution of strange quarks to the nucleon structure is of special interest because it is exclusively part of the quark-antiquark sea.

Recently, parity-violating electron scattering (PVES) experiments were used to probe the contribution of strange quarks to the structure of the nucleon. The strange quark content of the form factors is determined by assuming charge symmetry and combining parity-violating asymmetries with measurements of the electric and magnetic form factors of the proton and neutron [1, 2]. The first experimental results from the SAMPLE collaboration indicated that the strangeness content of both the magnetic moment and the radius of the proton was rather large and positive [3, 4], an unexpected and surprising finding, since a majority of theoretical studies favored a negative value for both quantities [2]. The situation has changed with the latest set of PVES experiments by the HAPPEX [5, 6, 7], PVA4 [8] and G0 [9] collaborations which showed much smaller values for the strange form factors. The most recent measurements show that the contribution of the strange quarks to the electric and magnetic form factors, G_E^s and G_M^s, is compatible with zero within the experimental errors.

Theoretically, it has been argued that the effects of strange quarks in the proton are not dominated by the lowest-lying virtual meson-baryon intermediate states, but rather are the result from cancellations between a large number of such intermediate states [10]. The aim of this contribution is to present a study of the contribution of strange quarks to the magnetic moment and the radius of the proton in an unquenched quark model in which the effects of the sea quarks are taken into account in an explicit way [11, 12].

2. Unquenched quark model

In the constituent quark model (CQM) hadrons are described as a system of constituent (or valence) quarks and antiquarks, qqq for baryons and $q\bar{q}$ for mesons. Despite the success of the quark model there is strong evidence for the existence of exotic degrees of freedom (other than valence quarks) in hadrons from CQM studies of electromagnetic and strong couplings of baryons that are on average underpredicted by CQMs [13, 14, 15, 16]. More direct evidence for the importance of quark-antiquark components in the proton comes from measurements of the $\bar{d}/\bar{u}$ asymmetry in the nucleon sea [17, 18], the proton spin crisis [19, 20] and PVES experiments [7, 8, 9].

The role of higher Fock components in baryon wave functions has been studied by many authors in the context of meson cloud models, the cloudy bag model, meson convolution models and chiral models [17, 21]. There have also been several attempts to study the importance of higher Fock components in the context of the constituent quark model. In this respect we mention the work by Riska and coworkers who introduce a small number of selected higher Fock components which are then fitted to reproduce the experimental data [22]. However, these studies lack an explicit model or mechanism for the mixing between the valence and sea quarks.

The impact of $q\bar{q}$ pairs in hadron spectroscopy was originally studied by Törnqvist and Zenczykowski in a quark model extended by the $^3P_0$ model [23]. Even though their model only includes a sum over ground state baryons and ground state mesons, the basic idea of the importance to carry out a sum over a complete set of intermediate states was proposed in there. Subsequently, the effects of hadron loops in mesons was studied by Geiger and Isgur in a flux-tube breaking model in which the $q\bar{q}$ pairs are created in the $^3P_0$ state with the quantum numbers of the vacuum [24, 25, 26]. In this approach, the quark potential model arises from an adiabatic approximation to the gluonic degrees of freedom embodied in the flux-tube [27]. It was shown that cancellations between apparently uncorrelated sets of intermediate states occur in such a way that the modification in the linear potential can be reabsorbed, after renormalization, in the new strength of the linear potential [25]. In addition, the quark-antiquark pairs do not destroy the good CQM results for the mesons [25] and preserve the OZI hierarchy [26] provided that the sum be carried out over a large tower of intermediate states. A first application of this procedure to baryons was presented in [10] in which the importance of $s\bar{s}$ loops in the proton were studied combining a $^3P_0$ pair-creation mechanism with harmonic oscillator wave functions for the baryons and mesons. This approach has the advantage that the effects of quark-antiquark pairs are introduced explicitly via a QCD-inspired pair-creation mechanism, which opens the possibility to study the importance of $q\bar{q}$ pairs in baryons and mesons in a systematic and unified way.

The present approach, called the unquenched constituent quark model (UCQM), is motivated by these earlier studies on extensions of the quark model to include the effects of $q\bar{q}$ pairs [10, 23]. Our approach is based on a CQM to which the quark-antiquark pairs with vacuum quantum numbers are added as a perturbation employing a $^3P_0$ model for the $q\bar{q}$ pair creation

[10, 11, 12]. The pair-creation mechanism is inserted at the quark level and the one-loop diagrams are calculated by summing over a complete set of intermediate baryon-meson states. Under these assumptions, the baryon wave function consists of a zeroth order three-quark configuration $\mid A\rangle$ plus a sum over all possible higher Fock components due to the creation of $^3P_0$ quark-antiquark pairs

$$\mid \psi_A\rangle \;=\; \mathcal{N}\left[\mid A\rangle + \sum_{BCl}\int d\vec{k}\mid BC\vec{k}\,l J\rangle\frac{\langle BC\vec{k}\,lJ\mid T^\dagger\mid A\rangle}{M_A - E_B - E_C}\right] . \qquad (1)$$

Here A denotes the initial baryon, B and C represent the intermediate baryon and meson, and M_A, E_B and E_C are their respective energies, $\vec{k}$ and l the relative radial momentum and orbital angular momentum of B and C, and J is the total angular momentum $\vec{J} = \vec{J}_B + \vec{J}_C + \vec{l}$. The operator $T^\dagger$ creates a quark-antiquark pair in the $^3P_0$ state with the quantum numbers of the vacuum: $L = S = 1$ and $J = 0$ [11, 12, 28]

$$T^\dagger \;=\; -3\gamma_0\sum_{ij}\int d\vec{p}_i\, d\vec{p}_j\, \delta(\vec{p}_i+\vec{p}_j)\, C_{ij}\, F_{ij}\, e^{-r_q^2(\vec{p}_i-\vec{p}_j)^2/6}$$
$$[\chi_{ij}\times\mathcal{Y}_1(\vec{p}_i-\vec{p}_j)]^{(0)}_0\; b_i^\dagger(\vec{p}_i)\, d_j^\dagger(\vec{p}_j)\; . \qquad (2)$$

Here, $b_i^\dagger(\vec{p}_i)$ and $d_j^\dagger(\vec{p}_j)$ are the creation operators for a quark and antiquark with momenta $\vec{p}_i$ and $\vec{p}_j$, respectively. The quark pair is characterized by a color singlet wave function C_{ij}, a flavor singlet wave function F_{ij} and a spin triplet wave function χ_{ij} with spin $S = 1$. The solid harmonic $\mathcal{Y}_1(\vec{p}_i - \vec{p}_j)$ indicates that the quark and antiquark are in a relative P wave. The operator $T^\dagger$ creates a pair of constituent quarks with an effective size, the pair creation point is smeared out by a Gaussian factor. The strong coupling vertex $\langle BC\vec{k}\,lJ \mid T^\dagger \mid A\rangle$ can be derived in explicit form in the harmonic oscillator basis [28, 29]. In the present calculations, we use harmonic oscillator wave functions with a single oscillator parameter for the baryons and another one for the mesons [30].

In order to calculate the effects of quark-antiquark pairs on an observable, one has to evaluate the contribution of all possible intermediate states. By using a combination of group theoretical and computational techniques, the sum over intermediate states is carried out up to saturation and not only for the first few shells as in previous studies [23, 10]. Not only does this have a significant impact on the numerical result, but it is necessary for consistency with the OZI-rule and the success of CQMs in hadron spectroscopy. In addition, the contributions of quark-antiquark pairs can be evaluated for any initial baryon (ground state or resonance) and for any flavor of the $q\bar{q}$ pair (not only $s\bar{s}$ as in [10], but also $u\bar{u}$ and $d\bar{d}$), and for any model of baryons and mesons, as long as their wave functions are expressed in the basis of harmonic oscillator wave functions [11, 12].

In the calculations presented in this contribution, we use harmonic oscillator wave functions up to four oscillator shells for the intermediate baryons and mesons. All parameters were taken from the literature without attempting to optimize their values in order to improve the agreement with experimental data [11, 12, 30].

3. Strange magnetic moment and strangeness radius

Even though the nucleon carries no net strangeness, it may have a nonvanishing distribution of strangeness. Kaplan and Manohar [31] observed that neutral current experiments could provide information on the strange matrix elements of the nucleon. Around the same time, it was suggested by Beck and McKeown [1] to use parity-violating electron scattering experiments to determine the strangeness contribution to the nucleon by combining the weak form factors

$G^{Z,p}$ of the proton with the electromagnetic form factors of the nucleon, $G^{\gamma,p}$ and $G^{\gamma,n}$. The strangeness contribution may then be extracted by performing a flavor decomposition [1, 2]

$$\begin{aligned} G^{\gamma,p}_{E,M}(Q^2) &= \frac{2}{3}G^u_{E,M} - \frac{1}{3}\left(G^d_{E,M} + G^s_{E,M}\right) , \\ G^{\gamma,n}_{E,M}(Q^2) &= \frac{2}{3}G^d_{E,M} - \frac{1}{3}\left(G^u_{E,M} + G^s_{E,M}\right) , \\ G^{Z,p}_{E,M}(Q^2) &= (1 - \frac{8}{3}\sin^2\theta_W)G^u_{E,M} + (-1 + \frac{4}{3}\sin^2\theta_W)\left(G^d_{E,M} + G^s_{E,M}\right) . \end{aligned} \tag{3}$$

The first measurements showed large and positive values of G^s_M, *e.g.* the SAMPLE collaboration in 1999 found $G^s_M = 0.61 \pm 0.17 \pm 0.21 \pm 0.19$ at $Q^2 = 0.1$ (GeV/c)^2 [3]. However, the more recent values for G^s_M and G^s_E obtained by the HAPPEX [6, 7], PVA4 [8] and G0 [9] collaborations are much smaller. The most recent measurements show that the contribution of strange quarks to the electric and magnetic form factors is compatible with zero within the experimental errors . Another method to determine the strangeness in the proton was suggested by Pate [32] which consists in combining experimental data on neutrino scattering with the electromagnetic form factors of the nucleon.

Theoretically, the strangeness of the proton has been addressed mostly in terms of two static observables, the strange magnetic moment μ_s and the strangeness radius r_s^2. The first theoretical calculation of the strange form factors of the nucleon performed by Jaffe in 1989 [33] reported quite large results for G^s_E and G^s_M which triggered a lot of interest in the contribution of strange quarks in the proton. Subsequent theoretical calculations of μ_s and r_s^2, obtained in lattice QCD calculations, hadronic models and effective hadronic theory, vary widely both in absolute value and sign [2, 34]. Fig. 1 shows a compilation of theoretical values of these two quantities. Most studies agree on a small negative strangeness radius and a moderately negative strange magnetic moment [2]. Recent lattice calculations give small negative values for both μ_s and r_s^2 [35, 36, 37] (see Tables 1 and 2).

The strange magnetic moment and the strangeness radius can be determined from the behavior of the strange form factors of the proton near the origin $Q^2 = 0$ as [2]

$$\begin{aligned} \mu_s &= G^s_M(0) , \\ r_s^2 &= -6 \left.\frac{dG^s_E}{dQ^2}\right|_{Q^2=0} . \end{aligned} \tag{4}$$

The values of μ_s and r_s^2 can be extracted from the experimental data as follows. In PVES experiments one determines a linear combination of the strange electric and magnetic form factors, $G^s_E + \eta G^s_M$ which can be separated by changing the scattering angle and/or the target (hydrogen, deuterium or ^{4}He).

Since the strangeness radius is proportional to the slope of G^s_E in the origin, the evaluation of r_s^2 requires a measurement of the strange electric form factor at a small value of Q^2 in combination with $G^s_E(0) = 0$ (the nucleon carries no net strangeness). Two global analyses of the available experimental data at low values of Q^2 give $r_s^2 = 0.014 \pm 0.096$ fm^2 [38] and $r_s^2 = 0.019 \pm 0.038$ fm^2 [39]. In Table 1 we show a compilation for the strangeness radius of the proton.

Since the strange magnetic moment is defined as $\mu_s = G^s_M(0)$, it involves an extrapolation of $G^s_M(Q^2)$ to $Q^2 = 0$. For example, in Ref. [4] the measured value of G^s_M at $Q^2 = 0.1$ (GeV/c)^2 is extrapolated by considering the momentum dependence of $G^s_M(Q^2)$ from [40]. The values obtained in a global analysis of the experimental data a low Q^2 are $\mu_s = 0.12 \pm 0.55 \pm 0.07$ μ_N [38] and $\mu_s = 0.29 \pm 0.21$ μ_N [39] (see Table 2).

Finally, we discuss the calculation of the strange magnetic moment and radius of the proton in the unquenched quark model. The results are obtained in a calculation involving a sum over

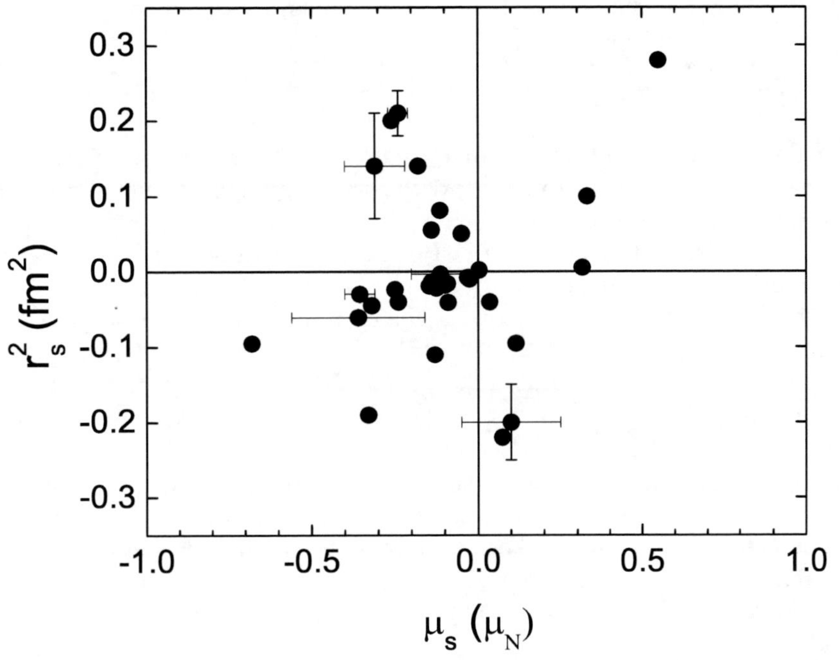

Figure 1. Theoretical values of the strange magnetic moment and the strangeness radius [2, 34]

Table 1. Strangeness radius of the proton

Reference	$r_s^2(p)$ [fm^2]
HAPPEX [6]	$0.098 \pm 0.108 \pm 0.026$
HAPPEX [7]	$-0.006 \pm 0.042 \pm 0.021$
HAPPEX [7]	0.012 ± 0.044
PVA4 [8]	$-0.053 \pm 0.040 \pm 0.020$
G0 [9]	$0.015 \pm 0.038 \pm 0.019 \pm 0.018$
Global Analysis [38]	0.014 ± 0.096
Global Analysis [39]	0.019 ± 0.038
LQCD [36]	$-0.007 \pm 0.004 \pm 0.002 \pm 0.021$
LQCD [37]	$-0.0024 \pm 0.0015 \pm 0.0007$
Unquenched QM [30]	0.012

Table 2. Strange magnetic moment of the proton

Reference	$\mu_s(p)$ [μ_N]
SAMPLE [4]	$0.01 \pm 0.29 \pm 0.31 \pm 0.07$
Global Analysis [38]	$0.12 \pm 0.55 \pm 0.07$
Global Analysis [39]	0.29 ± 0.21
LQCD [35]	-0.046 ± 0.019
LQCD [37]	$-0.017 \pm 0.025 \pm 0.007$
Unquenched QM [30]	0.0018

intermediate states up to four oscillator shells for both baryons and mesons. In the UCQM formalism, the strange magnetic moment of the proton is given by the expectation value of the operator

$$\vec{\mu}_s = \frac{1}{e_s} \sum_i \mu_{i,s} \left[2\vec{s}(q_{i,s}) + \vec{\ell}(q_{i,s}) - 2\vec{s}(\bar{q}_{i,s}) - \vec{\ell}(\bar{q}_{i,s}) \right] \; . \tag{5}$$

Here $\mu_{i,s}$ is the magnetic moment of the strange quark i (not be confused with the strange magnetic moment of the proton μ_s). In the UCQM the strange magnetic moment of the proton arises from the sea quarks. There is a contribution from the quark spins of the $s\bar{s}$ pair 0.0012 μ_N, as well as from its orbital motion 0.0006 μ_N. Both contributions are small and give a total strange magnetic moment $\mu_s = 0.0018\ \mu_N$ [30].

Similarly, the strange radius of the proton is calculated as the expectation value of the operator

$$r_s^2 = \sum_{i=1}^{5} \left(\vec{r}_{i,s} - \vec{R}_{CM} \right)^2 \; , \tag{6}$$

where $\vec{r}_{i,s}$ and $\vec{R}_{CM}$ are the coordinates of the strange quark i and the center of mass, respectively. The strangeness radius of the proton is calculated to be $r_s^2 = 0.012$ fm^2 [30].

In conclusion, the effects of the higher Fock components on the strange magnetic moment and the strange radius of the proton are found to be negligible. Tables 1 and 2 show that the results for the unquenched quark model are compatible with the latest experimental data and recent lattice calculations.

4. Summary and conclusions

In this contribution, we presented a study of the strangeness of the proton via two static observables, the strange magnetic moment and the strangeness radius, in the framework of an unquenched quark model for baryons in which the effects of sea quarks are taken into account in an explicit form via a $^3P_0$ creation mechanism of the quark-antiquark pairs ($u\bar{u}$, $d\bar{d}$ and $s\bar{s}$). The contribution of $s\bar{s}$ pairs to the magnetic moment and the radius of the proton was found to be small, in agreement with the latest experimental results and recent lattice calculations.

The results of the unquenched quark model obtained for the magnetic moments, the spin and flavor content of octet baryons [11, 12] and the strangeness of the proton [30] are very promising and encouraging. The inclusion of the effects of quark-antiquark pairs in a general and consistent way, as in the UCQM, may provide a major improvement to the constituent quark model which increases considerably its range of applicability.

Acknowledgments

This work was supported in part by CONACyT (grant 78833) and PAPIIT-UNAM (grant IN113711), Mexico and INFN, Italy.

References

[1] McKeown R D 1989 Phys. Lett. B **219** 140
Beck D H 1989 *Phys. Rev.* D **39** 3248
[2] Beck D H and McKeown R D 2001 *Ann. Rev. Nucl. Part. Sci.* **51** 189
Beck D H and Holstein B R 2001 *Int. J. Mod. Phys.* E **10** 1
[3] Spayde D T *et al* 2000 *Phys. Rev. Lett.* **84** 1106
Spayde D T *et al* 2004 *Phys. Lett.* B **583** 79
[4] Hasty R *et al* 2000 Science **290** 2117
[5] Aniol K A *et al* 1999 *Phys. Rev. Lett.* **82** 1096
Aniol K A *et al* 2001 *Phys. Lett.* B **509** 211
Aniol K A *et al* 2004 *Phys. Rev.* C **69** 065501
[6] K. A. Aniol *et al* [HAPPEX Coll.], Phys. Rev. Lett. **96** (2006) 022003.

[7] Acha A *et al* 2007 *Phys. Rev. Lett.* **98** 032301
[8] Baunack S *et al* 2009 *Phys. Rev. Lett.* **102** 151803
[9] Androic D *et al* 2010 *Phys. Rev. Lett.* **104** 012001
See http://www.npl.uiuc.edu/exp/G0/Backward
[10] Geiger P and Isgur N 1997 *Phys. Rev.* D **55** 299
[11] Bijker R and Santopinto E 2009 *Phys. Rev.* C **80** 065210
[12] Santopinto E and Bijker R 2010 *Phys. Rev.* C **82** 062202(R)
[13] Capstick S and Roberts W 1994 *Phys. Rev.* D **49** 4570
[14] Bijker R, Iachello F and Leviatan A 1994 *Ann. Phys. (N.Y.)* **236** 69
Bijker R, Iachello F and Leviatan A 1996 *Phys. Rev.* C **54** 1935
[15] Bijker R, Iachello F and Leviatan A 1997 *Phys. Rev.* D **55** 2862
Bijker R, Iachello F and Leviatan A 2000 *Ann. Phys. (N.Y.)* **284** 89
[16] Aznauryan I G *et al* 2008 *Phys. Rev.* C **78** 045209
Aznauryan I G *et al* 2009 *Phys. Rev.* C **80** 055203
Aznauryan I G, Burkert V D, Lee T S H and Mokeev V 2011 *J. Phys.: Conf. Ser.* **299** 012008
[17] Kumano S 1998 *Phys. Rep.* **303** 183
[18] Garvey G T and Peng J C 2001 *Prog. Part. Nucl. Phys.* **47** 203
[19] Ashman J *et al* 1988 *Phys. Lett.* B **206** 364
[20] Bass S D 2008 *The spin structure of the proton* (Singarpore: World Scientific)
Thomas A W 2009 *Int. J. Mod. Phys.* E **18** 1116
Kuhn S E, Chen J P and Leader E 2009 *Progr. Part. Nucl. Phys.* **63** 1
[21] Speth J and Thomas A W 1998 *Adv. Nucl. Phys.* **24** 83
[22] Zou B S and Riska D O 2005 *Phys. Rev. Lett.* **95** 072001
An C S, Riska D O and Zou B S 2006 *Phys. Rev.* C **73** 035207
Riska D O and Zou B S 2006 *Phys. Lett.* B **636** 265
Li Q B and Riska D O 2007 *Nucl. Phys.* A **791** 406
[23] Törnqvist N A and Zenczykowski P 1984 *Phys. Rev.* D **29** 2139
Törnqvist N A 1985 *Acta Phys. Polon.* B **16** 503 and 683
Zenczykowski P 1986 *Ann. Phys. (N.Y.)* **169** 453
[24] Kokoski R and Isgur N 1987 *Phys. Rev.* D **35** 907
[25] Geiger P and Isgur N 1990 *Phys. Rev.* D **41** 1595
[26] Geiger P and Isgur N 1991 *Phys. Rev. Lett.* **67** 1066
Geiger P and Isgur N 1991 *Phys. Rev.* D **44** 799
Geiger P and Isgur N 1993 *Phys. Rev.* D **47**, 5050
[27] Isgur N and Paton J 1983 *Phys. Lett.* B **124** 247
Isgur N and Paton J 1985 *Phys. Rev.* D **31** 2910
[28] Roberts W and Silvestre-Brac B 1992 *Few-Body Systems* **11** 171
[29] Ferretti J 2011 *Ph. D. Thesis* University of Genova, unpublished
[30] Bijker R, Ferretti J and Santopinto E 2012 *Phys. Rev.* C **85** 035204
[31] Kaplan D B and Manohar A 1988 *Nucl. Phys.* B **310** 527
[32] Pate S F 2004 *Phys. Rev. Lett.* **92** 082002
Pate S F, McKee D W and Papavassiliou V 2008 *Phys. Rev.* C **78** 015207
[33] Jaffe R L 1989 *Phys. Lett.* B **229** 275
[34] Bijker R 2006 *J. Phys. G: Nucl. Part. Phys.* **32** L49
[35] Leinweber D B *et al* 2005 *Phys. Rev. Lett.* **94** 212001
[36] Leinweber D B *et al* 2006 *Phys. Rev. Lett.* **97** 022001
[37] Doi T *et al* 2009 *Phys. Rev.* D **80** 094503
[38] Young R D *et al* 2006 *Phys. Rev. Lett.* **97** 102002
[39] Liu J *et al* 2007 *Phys. Rev.* C **76** 025202
[40] Hemmert T R, Meissner U G and Steininger S 1998 *Phys. Lett.* B **437** 184

Journal of Physics: Conference Series **387** (2012) 012012 doi:10.1088/1742-6596/387/1/012012

A panorama of excited-state g-factor measurements: advancing moment measurements on radioactive beams

Andrew E. Stuchbery

Department of Nuclear Physics, Research School of Physics and Engineering, The Australian National University, Canberra, ACT 0200, Australia

E-mail: `andrew.stuchbery@anu.edu.au`

Abstract. Patterns in the systematics of g factors for the first excited states of even-even nuclei are explored across the nuclear landscape for heavy nuclei ($A > 90$). The pathway to collectivity from semimagic to mid-shell nuclei is discussed, and attention given to the different behavior of nuclei either side of $N = 82$. As most $g(2_1^+)$ data is for nuclides near the valley of stability, the development of the recoil in vacuum technique for moment measurements on radioactive beams is discussed,with particular attention given to the free-ion hyperfine fields of Se and Ge ions leaving carbon foils with velocities of about 5% of the speed of light.

1. Introduction

In the first part of this paper, the gyromagnetic ratio systematics for the first-excited states in even-even nuclei, $g(2_1^+)$, are examined across the nuclear chart. The focus is on overall trends as well as the path from spherical to deformed nuclei and the role of g factors as a microscopic probe of the onset of collectivity. The evolution of nuclear structure is often discussed in terms of the ratio $R_{42} = E(4_1^+)/E(2_1^+)$ (see for example [1] and references therein). Figure 1 shows a nuclear chart with R_{42} values indicated by color, along the lines of Cakirli and Casten [2], but restricted to nuclides with $10 \leq Z \leq 82$ for which $g(2_1^+)$ is known. It is apparent that the $g(2_1^+)$ data are much less extensive than R_{42} data, and mainly cluster around the valley of stability. The second part of this paper therefore concerns the developing opportunities to perform g-factor measurements in new regions of the nuclear chart. The focus is on g-factor measurements on radioactive beams by the recoil-in-vacuum (RIV) technique [3, 4].

2. A panorama of excited state g factors in heavy nuclei: experimental trends

The trends in $g(2_1^+)$ values for medium to heavy nuclei have been discussed previously. It is timely, however, to reconsider these systematics in the light of theoretical advances (see below) and the availability of new data, which either increase the span of the data set (e.g. [5]) or significantly improve the precision (e.g. [6]). Discussion here will be on the three regions labeled in figure 1. It is convenient to begin with the heaviest region bounded by the $50 \leq Z \leq 82$ and $82 \leq N \leq 126$ shells, which will be referred to as the 'rare earth region'. The second region, the '$A \sim 130$' region, includes nuclei near ^{132}Sn having $Z > 50$ and $N \leq 82$. Finally, the '$A \sim 100$' region has nuclei with $40 \leq Z < 50$ and $50 \leq N < 80$.

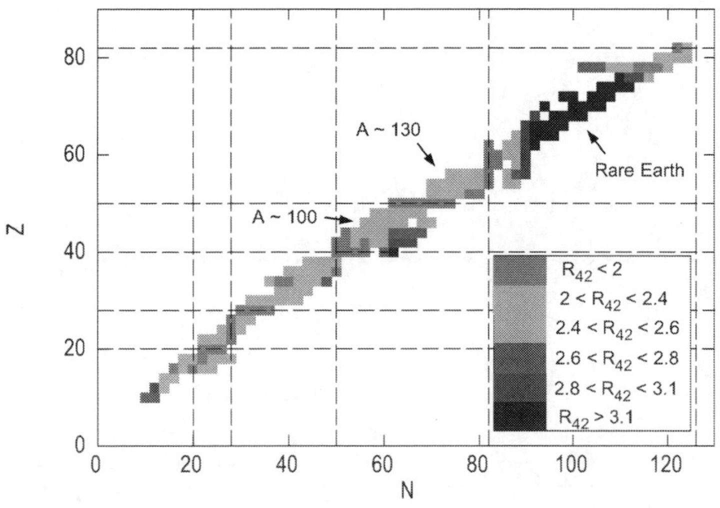

Figure 1. Chart of the nuclides showing nuclei between $Z = 10$ and $Z = 82$ for which $g(2_1^+)$ has been measured. The color indicates $R_{42} = E(4_1^+)/E(2_1^+)$ for each nuclide. Dashed lines indicate magic numbers.

To expose and investigate trends in nuclear moments for heavy nuclei, plots were made of $g(2_1^+)$ versus alternative variables such as R_{42}, N, and the valence proton fraction, N_p/N_t, where N_p is the number of valence protons or proton holes relative to the nearest magic number, and N_t is the total number of valence nucleons (protons plus neutrons) counted in the same way. Patterns and contrasting behavior proved most apparent when $g(2_1^+)$ was plotted versus N in the $A \sim 100$ region (see figure 2), and versus N_p/N_t in the $A \sim 130$ and rare-earth regions (see figures 3 and 4, respectively). Color has been used to indicate the R_{42} value associated with each measured g factor.

Looking first at the rare-earth region ($50 \leq Z < 82$ and $82 \leq N < 126$) and figure 2, it can be seen that the data for $N_p/N_t < 0.7$ cluster around a straight line, which can be interpreted in terms of the proton-neutron Interaction Boson Model expression $g = g_\nu + (g_\pi - g_\nu)(N_p/N_t)$. Although the boson g factors from the fit, $g_\nu = 0.23$ (neutrons) and $g_\pi = 0.45$ (protons), depart significantly from the nominal boson g factors, $g_\nu = 0$ and $g_\pi = 1$, the trend for an increase in $g(2_1^+)$ as the shell closures at $N = 82$ and $N = 126$ are approached is correct; in contrast, this experimental feature of the rare-earth region is contrary to the Z/A estimate, even when corrections for pairing are included [7].

It is known from observables such as R_{42} that heavy nuclei make a rapid transition from spherical to deformed behavior. The rapidity of this transition is evident in the g factors of the rare-earth nuclei as well. Aside from the semimagic $N = 82$ isotones, the nuclei that show strong departures from the collective trend line all have $R_{42} < 2.4$, and 4 or fewer valence neutrons. For the $N = 82$ nuclei $g(2^+) \sim 1$, the specific value being determined by the particular proton configuration. Then, with the addition of two neutrons, $g(2_1^+)$ falls well below the collective trend.

As shown in figure 3, in the $A \sim 130$ region the data for isotopes with $N_p/N_t < 0.7$ again cluster around a straight line, here characterized by $g_\nu = 0.31$ and $g_\pi = 0.38$. In contrast with the rare-earth region, there are no cases near $N_p/N_t = 0.8$ where the g factors undershoot the collective trend. The reason for this difference will be discussed below.

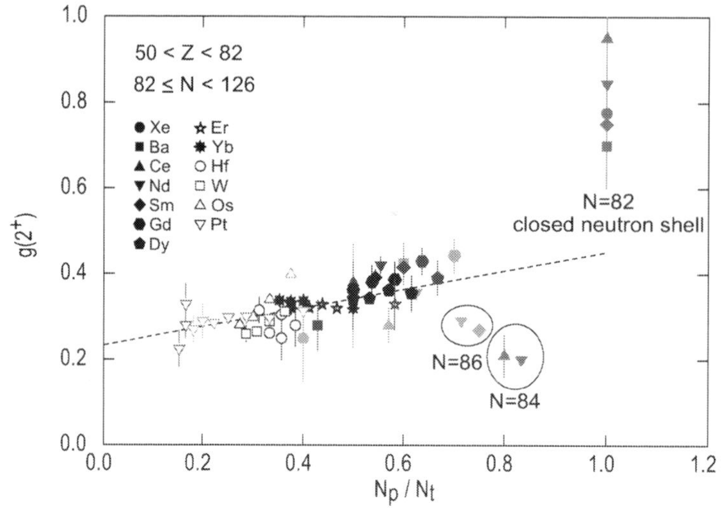

Figure 2. $g(2_1^+)$ as a function of the valence proton fraction for nuclei with $50 \leq Z < 82$ and $82 \leq N < 126$. The broken line indicates the trend for $N_p/N_t < 0.7$. Data points (from [8]) are colored to indicate R_{42}. See figure 1 for color key.

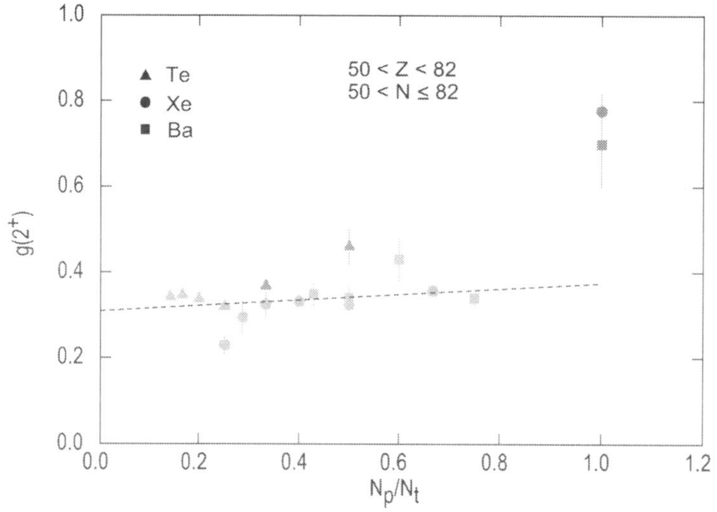

Figure 3. Similar to figure 2 but for nuclei with $50 < Z < 82$ and $50 < N \leq 82$.

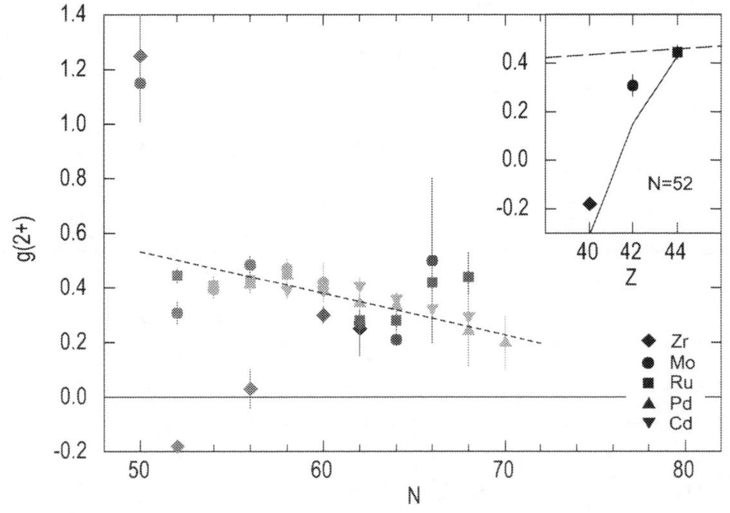

Figure 4. $g(2_1^+)$ as a function of neutron number for nuclei with $40 \leq Z < 50$ and $50 \leq N < 82$. Data are summarized in [6]. The broken line indicates the trend for $56 \leq N \leq 70$. See figure 1 for color key. The inset shows $g(2_1^+)$ for the $N = 52$ isotones. The dashed line is Z/A; the solid line is the shell model calculation of Holt *et al.* [9].

Finally, turning to figure 4 and the $A \sim 100$ region ($40 \leq Z < 50$ and $50 \leq N < 82$), the g factors departing strongly from the trend for collective nuclides show similarities to the rare-earth region in that the proton excitations in semimagic nuclei again have $g(2_1^+) \sim 1$, and then with the addition of two neutrons, $g(2_1^+)$ falls well below the collective trend, even becoming negative in ^{92}Zr.

Whereas the behavior of $g(2_1^+)$ follows a similar path to collectivity in the $A \sim 100$ and rare earth regions, the trends differ once collectivity sets in. In the $A \sim 100$ nuclei the g factors are *not* well correlated with N_p/N_t. Instead they decrease steadily as N increases beyond mid-shell at $N = 66$. The reason for this difference will be discussed below.

3. g factors and the onset of collectivity

The phenomenon whereby $g(2_1^+)$ values drop below the collective value of $g \sim Z/A$ in nuclei with two or four neutrons outside a closed shell was noted and discussed about a decade ago [10, 11]. An important observation is that the coupling between protons and neutrons in these nuclei with few valence nucleons is rather weak. For example, the level spectrum of $^{144}_{60}\mathrm{Nd}_{84}$ can be obtained by superimposing the level schemes of $^{142}_{60}\mathrm{Nd}_{82}$ (the 4-proton excitation) and $^{148}_{64}\mathrm{Gd}_{84}$ (the 2-neutron excitation; $^{142}_{64}\mathrm{Gd}_{82}$ is approximately a closed-shell nucleus), as shown in figure 5. It was suggested that (i) coupling between the valence protons and neutrons is relatively weak because the protons and neutrons are in different shells, and (ii) neutron excitations are then favored in the lowest 2^+ state because their residual interactions are more attractive than those between protons. Considerable progress has been made since then. The shell model calculations of Holt *et al.* [9] for the $N = 52$ isotones, shown in the insert of figure 4, give a good description of the data and illuminate the underlying nuclear structure phenomena. The weak coupling of the proton and neutron excitations leads to a 'configurational isospin polarization', which means that the 2_1^+ and 2_2^+ states in ^{92}Zr, for example, do not form the fully-symmetric and mixed-symmetry states of the proton-neutron interacting boson model. The consequent difference in

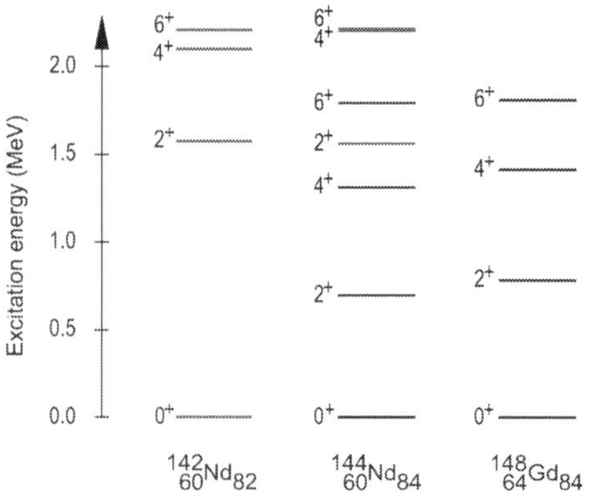

Figure 5. Level schemes demonstrating the weak coupling of proton (red) and neutron (blue) neutron excitations in ^{144}Nd.

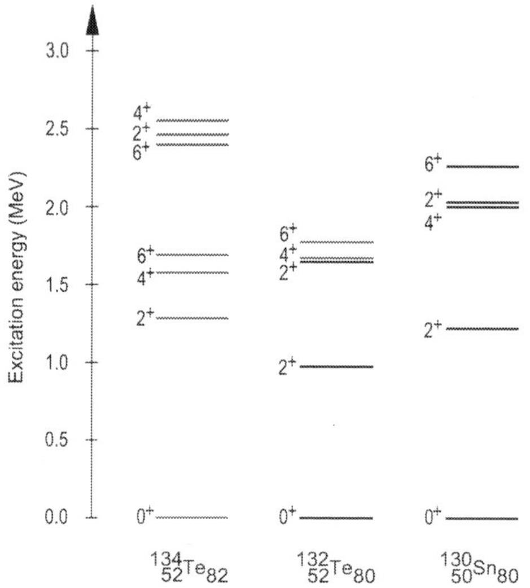

Figure 6. Similar to figure 5 but showing the similar energies of the parent proton and neutron excitations in the lowest 2^+ state of ^{132}Te.

their g factors has been observed [12].

Research into the g factors of nuclei near closed shells, the formation of mixed symmetry states, and the on-set of collectivity, is on-going. In the $A \sim 80$ region, the $N = 48$ isotones $^{84}_{36}$Kr and $^{86}_{38}$Sr follow the pattern of $g(2^+_1) \ll Z/A$. On the other hand nuclei in the $A \sim 130$ region do not follow the same trend. For example, $^{132}_{54}\mathrm{Te}_{80}$, which has two valence protons and two neutron holes, has $g \sim Z/A$ [3, 4]. This case has been discussed recently by Danchev *et al.* [13]; the 2^+_2 state was populated by Coulomb excitation of a radioactive beam and identified as the first example of a mixed symmetry state in a neutron-rich nucleus.

Further insight into the different behavior of the $A \sim 130$ nuclei is gained from figure 6 wherein the level scheme of $^{132}_{52}\mathrm{Te}_{80}$ is compared with those of $^{130}_{50}\mathrm{Sn}_{80}$ and $^{134}_{52}\mathrm{Te}_{82}$, which represent its constituent neutron and proton configurations, respectively. It is evident that the 2^+_1 proton excitation in ^{134}Te is nearly degenerate with the 2^+_1 neutron excitation in ^{130}Sn, and as a consequence the 2^+_1 and 2^+_2 states in ^{132}Te have about equally mixed proton and neutron components. This scenario is confirmed by shell model calculations [13]. Looking to the 4^+_1 and 6^+_1 states, it is evident from the comparison of the levels in figure 6 that, in contrast to the 2^+ states, these states in ^{132}Te are likely to be predominantly proton excitations akin to the 4^+_1 and 6^+_1 states in ^{134}Te. The measured g factors of the 6^+_1 states [8] are $g(^{134}\mathrm{Te}) = +0.847(25)$ and $g(^{132}\mathrm{Te}) = +0.78(8)$, giving further evidence that both states are essentially the same proton excitation.

A comparison of the level schemes of $^{130}_{52}\mathrm{Te}_{78}$, $^{128}_{50}\mathrm{Sn}_{78}$ and $^{134}_{52}\mathrm{Te}_{82}$ shows that with the removal of a further two neutrons, the relative level ordering remains very like the case of ^{132}Te. It is to be expected, therefore, that in ^{130}Te the lowest two 2^+ states are mixed proton-neutron excitations, whereas the 4^+_1 state is predominantly a proton excitation. An experiment is planned at the Australian National University to measure these three g factors in ^{130}Te simultaneously by the transient-field method [14]. As more comprehensive data are obtained concerning the g factors of excited states near closed shells, it will help develop deeper insights into the formation of mixed symmetry states and the on-set of collectivity in atomic nuclei.

4. Global trends in collective g factors

As noted above, the behavior of $g(2_1^+)$ follows a similar path to collectivity in the $A \sim 100$ and rare earth regions, but the trends differ once collectivity sets in. Why the difference? It must stem from the sensitivity of the g factors to the underlying single-particle composition of the quadrupole collectivity.

The question has been illuminated by microscopic calculations using the tidal-wave model of Frauendorf and collaborators [6]. This model, which describes the yrast states of transitional and deformed nuclei by means of the self-consistent cranking model, allows the calculation of the magnetic moment directly from the nucleonic currents. It was found that the decrease in $g(2_1^+)$ along the isotope chains in the $A \sim 100$ region is primarily due to an increasing angular momentum contribution from the $h_{11/2}$ neutrons, which can begin even below $N = 64$, as illustrated for the Mo isotopes in figure 7. The mechanism is akin to the strong increase of angular momentum carried by neutrons in deformed nuclei, caused by the rotational alignment of the $i_{13/2}$ and $j_{15/2}$ neutrons, which reduces the g factors of high-spin states below Z/A.

It is therefore proposed that the difference in the g-factor systematics observed for the $N = 50 - 82$ shell compared to the $N = 82 - 126$ shell stems from the difference in the location of the high-spin intruder orbital. As seen in figure 7, the neutron $h_{11/2}$ orbit is near the top ($N = 82$ end) of the $N = 50 - 82$ shell. In contrast, the $i_{13/2}$ orbit is much nearer to the middle of the $N = 82 - 126$ shell. A study of the data in figure 2 in the tidal-wave model is needed. The aim would be to provide a microscopic basis for the trends observed, including the trend of rising g factors towards the $N = 126$ end of the shell.

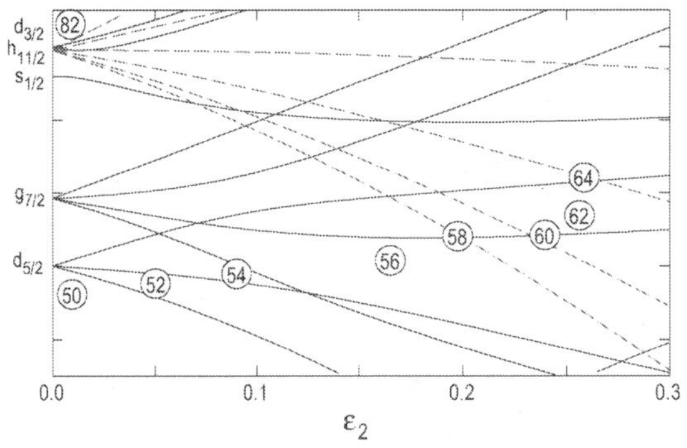

Figure 7. Nilsson diagram for neutrons in Mo isotopes, which illustrates the intrusion of the $h_{11/2}$ shell among the positive parity orbits and the consequent impact of $h_{11/2}$ orbit on the g factors in this region [6]. Neutron numbers are indicated at the calculated deformations.

5. Moment measurements on radioactive beams: the Recoil in Vacuum method

It is apparent from the preceding discussion that g-factor measurements can give useful insights into nuclear structure, both to understand global trends and the details of specific regions. However, as indicated in figure 1, $g(2_1^+)$ measurements are at present largely confined to stable nuclides. This limitation is being overcome through measurements on radioactive beams. In this section attention turns to some developments related to the recoil in vacuum method.

When a free ion moves through vacuum, the hyperfine interaction couples the atomic spin $\boldsymbol{J}$ to the nuclear spin $\boldsymbol{I}$ and together they precess about the total spin $\boldsymbol{F} = \boldsymbol{I} + \boldsymbol{J}$. The precession frequency $\omega_{F,F'}$ is proportional to the nuclear g factor and the magnitude of the hyperfine magnetic field at the nucleus. To measure the g factor, the nuclear state of interest is excited by a suitable reaction and then allowed to recoil into vacuum. The effect of the hyperfine interaction is observed via the perturbation of the angular correlation of the γ-rays de-exciting the state. The difference between the perturbed angular correlation and the unperturbed angular correlation is described by the vacuum attenuation coefficients, G_k ($k = 2, 4$), which also contain the information about the nuclear g factor. This recoil in vacuum (RIV) method has proved suitable to measure excited-state g factors in unstable nuclei, especially neutron-rich nuclei produced as radioactive beams [3, 4]. For further details of the method and notation, see [4] and references therein.

The RIV method has a number of advantages for applications to radioactive beams, one being that both the g factor and the $B(E2; 0 \to 2)$ can be determined in the same experiment. At the Australian National University we have been studying the free-ion hyperfine fields of stable nuclei having known moments with a view to the applications of the RIV technique to radioactive beams. These free-ion fields must be characterized in order to extract the nuclear g factor from the measured attenuation coefficients, G_k. In contrast to the radioactive beam measurements, which use large arrays of particle and γ-ray detectors with large solid angle coverage, the stable beam measurements have been made with modest apparatus. Four HPGe γ-ray detectors operate in coincidence with an array of eight photodiode particle detectors called 'Heliotrope' placed around the beam axis at forward angles ($\sim 30°$). Like the radioactive beam measurements, the stable-beam studies have been performed in inverse kinematics with the beam ions Coulomb excited on either ^{12}C or ^{27}Al targets; the different targets serve to vary the exit velocity and hence charge-states of the ions entering vacuum.

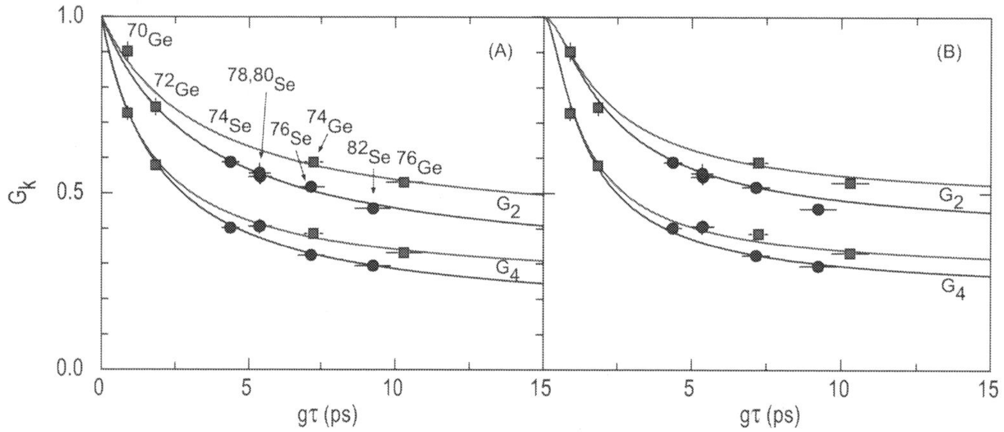

Figure 8. Vacuum deorientation coefficients for the stable isotopes of Ge and Se after excitation of 175 MeV beams on carbon targets. (A) Solid lines are empirical fits to equation 1. (B) Lines are model-based fits assuming gaussian field distributions (see text).

Figure 8 shows measured attenuation coefficients, G_k, for 175 MeV beams of the stable $_{32}$Ge and $_{34}$Se isotopes excited on a carbon target. As in previous work, the G_k values are plotted as a function of the product of the g factor and the mean life of the 2_1^+ states. The observed difference in the attenuation for the longer-lived isotopes, i.e. ^{74,76}Ge compared with ^{76,82}Se, was surprising. The measurements were therefore repeated and subjected to extensive checking. For example, a cocktail beam of ^{74}Ge and ^{74}Se was used to perform simultaneous measurements of the vacuum attenuation for these two ions.

We do not yet have a 'first-principles' atomic physics explanation for this difference between the hyperfine fields for Ge and Se ions recoiling in vacuum under almost identical conditions. However some important conclusions can be drawn based on empirical and semi-empirical fits to the data. The left panel in figure 8 shows an empirical fit based on

$$G_k = \alpha_k + (1 - \alpha_k)\frac{C_k}{C_k + g\tau}, \tag{1}$$

Journal of Physics: Conference Series **387** (2012) 012012 doi:10.1088/1742-6596/387/1/012012

where α_k and C_k are parameters that must be fitted separately to G_2 and G_4. The 'hard core' parameter α_k gives the asymptotic value of G_k at long times, while C_k is the time constant for the quasi-exponential decay of the attenuation coefficient. Physically, C_k is related to the average strength of the hyperfine fields acting on the nucleus, while α_k is determined by the average angular momentum of the atomic electron configurations.

The right panel of figure 8 shows a semi-empirical model-based fit to the data. This static-model fit method has been described in [4]. It assumes that the vacuum attenuation results from the superposition of hyperfine fields with gaussian distributions in atomic spin and field strength. To limit the parameters it has proved effective to (i) fix the standard deviation of the atomic spin to $\sigma_J = 1\hbar$, and (ii) for the distribution of hyperfine fields, to set $\sigma_B = \bar{B}$, where $\bar{B}$ is the average hyperfine field strength at the nucleus. Two parameters, $\bar{J}$ and $\bar{B}$, then determine G_2 and G_4.

Both the empirical fit (with 4 parameters) and the semi-empirical model-based fit (with 2 parameters) give comparable descriptions of the data. Moreover, both fits show that effectively the *same* average hyperfine field strength is experienced by the Ge and Se ions. What differs for the Ge versus Se ions is the magnitude of the hard core term in the empirical fit or, equivalently, the value of $\bar{J}$ in the model-based fit. Specifically, $\bar{J} = 1.2$ for Ge and $\bar{J} = 1.7$ for Se.

Charge-state measurements (performed at the Australian National University) suggest that the Ge ions here carry on average 12 to 13 electrons, whereas the Se ions on average carry 15. The Ge ions are therefore mainly Mg-like and Al-like, with ground state configurations of $3s^2$ and $3s^2 3p^1$, respectively. These configurations produce the terms ^1S and ^2D, respectively, with corresponding ground-state atomic spins of $J = 0$ and $J = 1/2$. For P-like ions the lowest configuration is $3s^2 3p^3$, which produces the terms ^4S, ^2D, and ^2P, and the ground-state spin is $J = 3/2$. These observations are qualitatively consistent with the experimental evidence that $\bar{J}$ is slightly higher for Ge than for Se. Also, the low-excitation spectra of excited Mg-, Al- and P-like ions are consistent with an average atomic spin in the range $1 \leq \bar{J} \leq 2$.

Figure 9 shows how the hard core parameters depend on the atomic angular momentum J. The magnitude of the hard-core is a very strong function of J for $J < 2$, but becomes almost independent of the atomic angular momentum for $J \geq 2$. It can be concluded that the strong

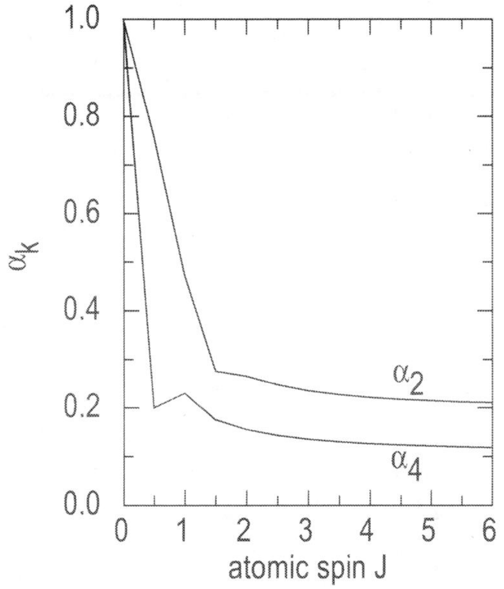

Figure 9. Atomic spin dependence of the hard-core vacuum-deorientation coefficients.

sensitivity of the hard-core attenuation observed for Se and Ge ions here is a consequence of their average atomic spin being less than 2 $\hbar$. In cases where $\bar{J} \geq 2$, less sensitivity to the atomic number and recoil velocity, both of which affect the average atomic configuration, is expected.

Further experiments are underway to investigate the velocity dependence of these hyperfine fields and perform detailed measurements of the relevant charge-state distributions. Calculations of the hyperfine fields for specific atomic configurations are being pursued to provide a more quantitative explanation of the free ion hyperfine fields of these highly charged ions. Due to their strong sensitivity to the average atomic angular momentum, the above data for Ge and Se ions provide a stringent test for 'first-principles' atomic calculations of free ion hyperfine fields (see for example, [15]).

6. Summary and Conclusion

Patterns in $g(2_1^+)$ systematics for heavy nuclei have been examined and found to be sensitive to the underlying single-particle structure. Near closed shells the magnetic moments can show pronounced changes that depend on the strength of the coupling between protons and neutrons. Beams of radioactive ions are enabling the extension of the $g(2_1^+)$ data to include unstable nuclei, especially neutron-rich isotopes.

The RIV method has proved useful for magnetic moment measurements on the first-excited states of unstable nuclei produced as radioactive beams. Experiments on stable beams with known moments are underway at the Australian National University to characterize the free ion hyperfine fields, which must be understood to fully exploit the RIV method. An unexpected difference between the hyperfine fields for Ge and Se ions recoiling in vacuum under similar conditions can be attributed to their average atomic angular momentum being between 1 and 2 $\hbar$. Such cases will provide a sensitive testing ground for atomic physics calculations that seek to calculate the hyperfine fields from first principles.

Acknowledgments

The author is grateful to many colleagues who have made the work discussed here possible. Russell Leslie in particular is thanked for his contributions to the measurements described in section 5. This research was supported in part by the ARC Discovery Scheme, Grant No. DP0773273.

References

[1] Cejnar P, Jolie J and Casten R F 2010 *Rev. Mod. Phys.* **82** 2155
[2] Cakirli R B and Casten R F 2006 *Phys. Rev. Lett.* **96** 132501
[3] Stone N J *et al* 2005 *Phys. Rev. Lett.* **94** 192501
[4] Stuchbery A E and Stone N J 2007 *Phys. Rev. C* **76** 034307
[5] Berant Z *et al* 2009 *Phys. Rev. C* **80** 057303
[6] Chamoli S K *et al* 2011 *Phys. Rev. C* **83** 054318
[7] Stuchbery A E 1995 *Nucl. Phys.* **A 589** 222
[8] Stone N J 2011 **INDC(NDS)-0594** http://www-nds.iaea.org/publications/indc/indc-nds-0594/
[9] Holt J D, Pietralla N, Holt J W, Kuo T T S and Rainovski G 2007 *Phys. Rev. C* **76** 034325
[10] Stuchbery A E 2001 *Nuclear Physics A* **682** 470
[11] Mantica P F, Stuchbery A E, Groh D E, Prisciandaro J I and Robinson M P 2001 *Phys. Rev. C* **63** 034312
[12] Werner V *et al* 2008 *Phys. Rev. C* **78** 031301
[13] Danchev M *et al* 2011 *Phys. Rev. C* **84** 061306
[14] Stuchbery A E *et al* 2007 *Phys. Rev. C* **76** 034306
[15] Stone N J, Stone J and Jonsson P 2010 *Hyperfine Interactions* **197** 29

Precise Coulomb excitation B(E2) measurements for first 2^+ states of projectile nuclei near the doubly magic nuclei ^{78}Ni and ^{132}Sn

A Galindo-Uribarri [1,2]
[1] Physics Division, Oak Ridge National Laboratory, Oak Ridge TN, 37931, USA
[2] University of Tennessee, Knoxville TN, 37996, USA

E-mail: uribarri@phy.ornl.gov

Abstract. Coulomb excitation is a very precise tool to measure excitation probabilities and provide insight on the collectivity of nuclear excitations and in particular on nuclear shapes. In the last few years radioactive ion beam facilities such as HRIBF opened unique opportunities to explore the structure of nuclei in the regions near the doubly magic nuclei ^{78}Ni (Z=28 and N=50) and ^{132}Sn (Z=50 and N=82). For this purpose we have developed specialized methods and instrumentation to measure various observables. There is also the opportunity to perform precision experiments with stable beams using exactly the same state-of-the-art instrumentation and techniques as with their radioactive ion beam counterpart. I describe some of the recent efforts at HRIBF to do more precise measurements using particle-gamma techniques.

1. Introduction

In the last few years various schemes to produce and accelerate nuclei far from stability have been developed, opening new and exciting opportunities to explore nuclear properties at extreme values of Z/N (with Z protons and N neutrons) and to better understand the forces that bind the nucleons together. In nuclear structure studies, the Radioactive Ion Beams (RIBs) can be considered as a tool that allows us to explore new regions in the nuclear chart and to study the shell structure and the gap evolution when the number of neutrons and protons changes providing a stringent test on the applicability of the theoretical models.

The use of RIBs for nuclear structure studies at low energy presents several experimental difficulties that require the development of methods and instrumentation. Each radioactive beam experiment presents its own set of problems and challenges due to the radioactive background from the beam, beam contaminants and the low beam intensity.

For many years, Coulomb excitation and transfer reactions have been used with stable beams and have provided a wealth of nuclear structure information. At the Holifield Radioactive Ion Beam Facility (HRIBF) at Oak Ridge National Laboratory we have developed specialized instrumentation and techniques for performing these reactions in inverse kinematics with unstable beams. In particular we are using these two powerful techniques for the study of nuclear structure at low energies using neutron-rich RIBs near closed shells (e.g. ^{78}Ni and ^{132}Sn). While the primary focus of the HRIBF is the study of nuclear reaction spectroscopy and nuclear astrophysics using RIBs, there is also the opportunity to perform precision experiments with stable beams. Access to good quality beams from the facility and the newly developed powerful instrumentation provides unique opportunities to perform systematic studies along isotopic chains using exactly the same instrumentation and techniques as with their RIB counterpart.

2. Coulomb Excitation and transfer reactions

Coulomb excitation is a purely electromagnetic excitation process of nuclear states due to the Coulomb field of two colliding nuclei. As such, it is a very precise tool to measure excitation probabilities and provide insight on the collectivity of nuclear excitations and in particular on nuclear shapes. At low beam energies, such as the ones we use at HRIBF, the interaction between the beam and the target nuclei is purely electromagnetic and predominantly populates the first excited state. Therefore, there is no need for corrections that account for feeding from the higher-lying levels, or potential nuclear-interference effects, as is the case of higher energy Coulomb excitation.

At HRIBF, we have made pioneer measurements of B(E2) values for several neutron-rich radioactive nuclei in the mass A = 80 and A =130 regions [1,2]. We have repeated some of these measurements with increased statistics [3]. Also we are revisiting some of the reported measurements of B(E2) and quadrupole moments using stable beams along chains of isotopes such as Ni, Ge, Se, Sr, Sn, and Te to resolve some existing discrepancies. The approach has been to do safe Coulomb excitation reactions between various projectile-target combinations that allow for consistency crosschecks, to keep the corrections to the data to a minimum (e.g. electronics deadtime), and to extract and compare the results obtained using various independent techniques (e.g. normalization to Rutherford vs. normalization relative to the target well known B(E2) values). Together with the results obtained with RIBs (and using the same techniques) these revisited data will allow for a more meaningful comparison and a stringent test to theoretical predictions.

Another very successful program at HRIBF has been the study of transfer reactions using particle-gamma techniques [4]. Transfer reactions give direct access to the microscopic shell structure of nuclei. One-nucleon "pickup" (p, d) and one-nucleon "stripping" (d, p) reactions are useful spectroscopic tools in inverse kinematics with RIBs. Their advantages include simplicity on the interpretation and large transfer cross-sections for single particle (or hole) states. Techniques for performing the above reactions in inverse kinematics with unstable beams are being developed however, substantial development is still required in both targets and detection systems to fully exploit the potential of transfer reactions in hydrogen isotopes. Alternative approaches include the use of heavy ion transfer reactions with RIBs which we are exploring at HRIBF as a technique to study the structure of nuclear states inaccessible or difficult to populate in light ion reactions. It is possible to reach the same final nucleus by using a wide variety of different projectiles and by transferring different numbers of nucleons. Much may then be deduced concerning the structure of a state simply from its population or non-population in a series of different reactions. An attractive feature of heavy ion reactions below the Coulomb barrier is their insensitivity to the nuclear interaction between the cores simplifying the analysis.

Another approach we are pursuing is to develop techniques that allow us to measure more sensitive observables. The use of a polarized target in transfer reactions for example would allow the measurement of observables such as the analyzing power which gives an unambiguous determination of the total spin J [5].

The use of RIBs to do both Coulomb excitation reactions and transfer reactions presents serious challenges associated to their low intensities, potentially high backgrounds in the gamma-ray and charged particle detectors caused by the radioactivity of the beam, and isobaric contamination. Some of these problems can be solved with specially designed detection systems, as described later, and with good quality beams.

3. Beam production

The RIBs are produced using the Isotope Separator On-Line (ISOL) approach. The ISOL method requires two accelerator systems, a driver accelerator to produce radioactive atoms at rest and a post accelerator to accelerate them to energies of interest. The two accelerators are coupled by a target-ion

source (TIS) and a high-resolution mass separator. Facilities that use the ISOL method include HRIBF at ORNL in USA, ISAC-TRIUMF in Canada, and ISOLDE at CERN. HRIBF has a great advantage over other ISOL facilities that it has access to excellent quality beams for many stable species in a wide range of energies.
At HRIBF (Figure 1), the K = 100 Oak Ridge Isochronous Cyclotron (ORIC) serves as the driver, accelerating intense beams (up to 20 μA) of light ions (p, d, alphas) for the production of radioactive isotopes, and the folded-geometry 25MV tandem accelerator serves as a post accelerator for the RIBs [6]. This accelerator is the highest operating voltage electrostatic accelerator in the world capable of producing beams from very low energies up to 15 MeV per nucleon for light nuclei and up to 5 MeV per nucleon for heavier masses. The ISOL method involves a thick target maintained at high temperatures (T ~ 2200 °C) to promote the rapid transfer to the actual ion source. Once ionized in an ion source the ions can be extracted by electric fields and then mass analyzed with magnets before its injection into a post-accelerator which takes them to the energy required by the experiment. HRIBF is the first facility to produce post-accelerated beams of heavy neutron-rich nuclei. In fact, the majority of the RIBs available at HRIBF are on the n-rich side (Figure 2). They originate from the fission spectra of ^{nat}U bombarded with protons of energy of about 50 MeV. The production target for n-rich nuclei consists of UC_2 mixed with graphite.

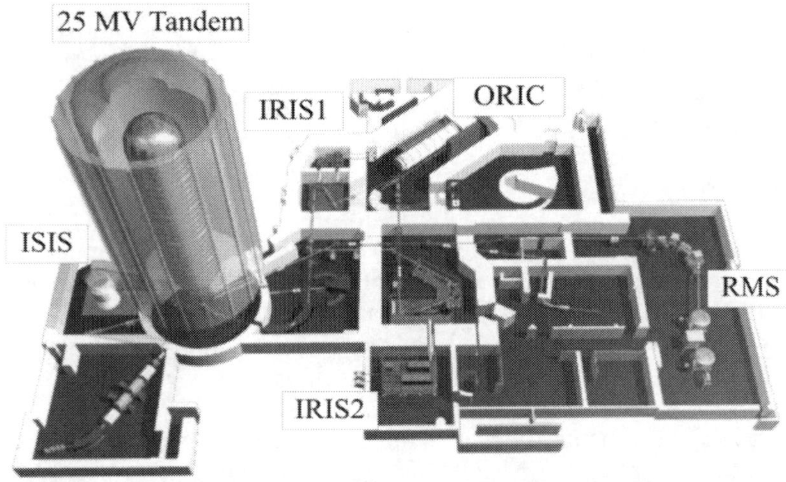

Figure 1. Schematic layout of HRIBF, an ISOL facility at Oak Ridge National Laboratory whose main components are a K=110 cyclotron (ORIC) serving as a "driver", a 25MV Tandem as a post-accelerator, two target/ion-sources (IRIS1 and IRIS2) for RIB production, an injector for stable ion species (ISIS) and various experimental stations. The majority of the Coulomb excitation experiments were performed at the RMS target station.

4. Particle-gamma detection: experimental considerations

A large solid angle particle-gamma combination has proven a very powerful tool in nuclear spectroscopy, because of the ability to choose the excitation energy and angular momentum, isolate different exit channels, identify charged particles, and measure their multiplicities, angular correlations, and energy spectra with very large efficiencies [7]. Using charged particle arrays, the momenta of the particles detected can be used to reconstruct the reaction kinematics in each event, thus enabling a considerable reduction of the Doppler broadening of the peaks of the γ-ray spectra [8]. Use of these arrays also greatly improves the reaction channel selectivity through energy-conservation constraints [9]. Important discoveries can be attributed directly to the use of 4π charged particle arrays

Journal of Physics: Conference Series **387** (2012) 012013 doi:10.1088/1742-6596/387/1/012013

in conjunction with large gamma spectrometers: (i) the discovery of band termination and collective states in ^{64}Zn[7]; (ii) the experimental and theoretical study of collective properties of ^{48}Cr at high spin [10-12]; superdeformation in the mass A=40 region in ^{36}Ar [13] and ^{40}Ca [14], to name a few.

At HRIBF, we have developed particle-gamma instrumentation and techniques to perform Coulomb excitation reactions and transfer reaction studies (Figure 3).

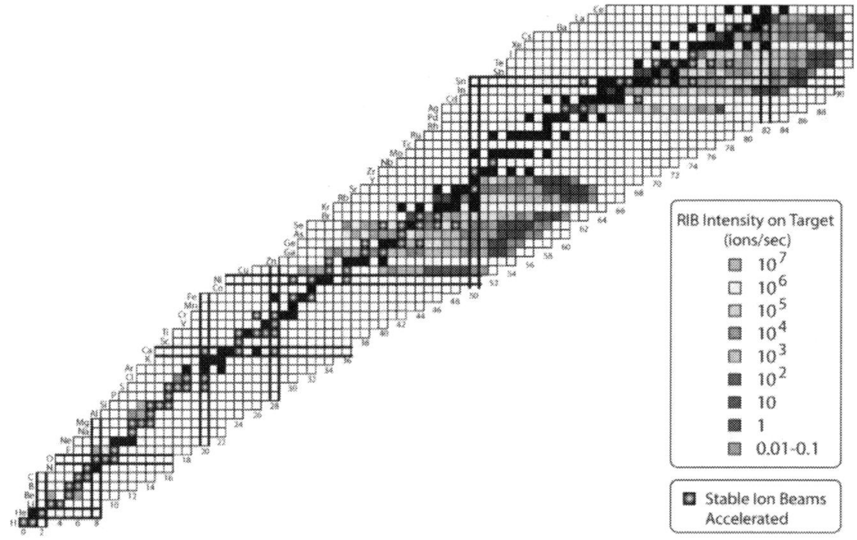

Figure 2. HRIBF Post-accelerated Beams. Section of the chart of nuclides showing intensities of 3 MeV/A beam-on-target of neutron-rich beams at the HRIBF. In addition to about 80 stable ion beam species, about 175 RIB species are available (32 proton-rich species and 143 neutron-rich species).

For Coulomb excitation experiments the gamma-rays emitted by the projectile and target were detected in a gamma-detector array (CLARION) and the target nuclei were detected with a charged particle array with minimum absorbers (BAREBALL). A fraction of the beam was sampled using an axial ionization chamber or Bragg curve detector placed at zero degrees to determine and monitor any change in the beam composition. Most importantly the Bragg detector was also used to measure beam energy loss in the target (Figure 4). This way we know precisely to better than 1% in a target in/out sequence the energy loss of the beam in the specific target used avoiding an important source of error. Tests have been performed with the Bragg detector to measure target thickness homogeneity and energy loss variations due to intense beam bombardment.

Figure 3. HYBALL array of CsI(Tl) crystals mounted inside the CLARION gamma-ray spectrometer. Each clover detector is surrounded by a BGO Compton-suppression shield.

Germanium, selenium, tin and tellurium isotopes have been Coulomb excited by bombarding light targets (e.g. ^{12}C, ^{27}Al, ^{24}Mg, and ^{50}Ti) of about 1 mg/cm^2 thickness using "safe" energies. Criteria to establish the maximum bombarding energy where the interaction between the nuclei is dominated by Coulomb forces ("safe" energy) is discussed in references [15-17].

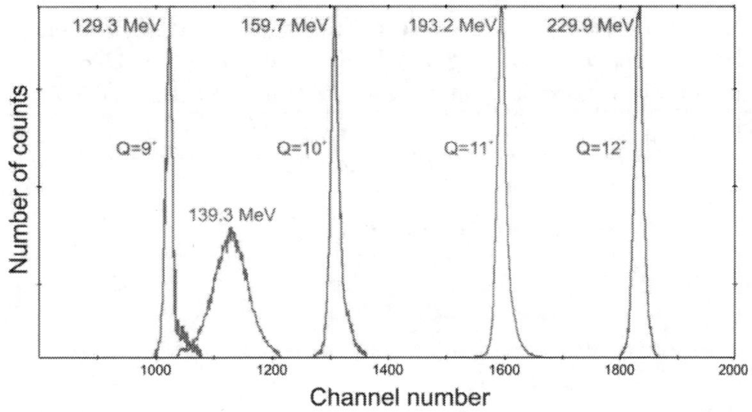

Figure 4. Superimposed Bragg curve detector energy spectra for various charge states. The broader peak at 139.2 MeV corresponds to the beam going through the target. We used a setup that allows target in/out sequence to measure the beam energy loss in the specific target used in the experiment. Different charge states of beam with same analyzing magnet settings were used to calibrate the Bragg counter response.

5. Precise B(E2) measurements using Coulomb excitation

The Coulomb excitation process has an advantage over other nuclear reactions in that the electromagnetic interaction is well understood and the theory of the interaction has been well developed. Under the proper "safe" conditions the projectile nucleus and the target are never close to each other for the nuclear force to play a role and the excitation is accomplished purely through Coulomb interaction. In principle it is possible using modern experimental techniques to determine in a model-independent way the M1, E1, E2 and E3 matrix elements connecting low lying collective nuclear levels populated by electromagnetic excitation. There are several prescriptions for achieving "safe" Coulomb excitation conditions but the best approach is to experimentally investigate them by bombarding at various energies. This is particularly important when the reaction partners involve deformed or very light nuclei.

Coulomb excitation has many advantages as a mechanism for excitation of nuclear states reaching very neutron-rich radioactive species, as the projectile itself can be excited. Large cross-sections are involved helping to achieve good statistics in the measured data even for weak RIBs. By contrast, very neutron-rich compound nuclei formed in fusion-evaporation reactions will tend to evaporate several neutrons bringing the residual nuclei towards the valley of stability.

Experimental B(E2) values for several of the isotopic chains in the mass A = 80 and A =130 regions have been measured with different techniques and experimental conditions, and show considerable spread (see compilation of B(E2) values in the "Raman Tables" [18]). I show an example in Figure 5 of the spread of B(E2) values for the N = 40 ^{72}Ge. It is important to determine these values precisely for some stable nuclei as some of these "well-known" B(E2) values have been used as a calibration measurement or reference for the determination of B(E2) values of radioactive species.

We have measured the B(E2) values of a series of radioactive beams of Ge, Se, Sn and Te isotopes [1,2,19]. Our approach has been to get a comprehensive picture of the shell structure in these regions by studying a series of properties of low lying states (E (2^+), B (E2), g-factors and quadrupole

Journal of Physics: Conference Series **387** (2012) 012013 doi:10.1088/1742-6596/387/1/012013

moments). The beams, instrumentation and techniques developed specifically for this purpose have allowed us to systematically study the behavior of these observables along isotopic and isotonic chains *using both stable and radioactive nuclei under almost identical experimental conditions.*

We have successfully developed a technique to Coulomb excite a RIB, in which scattered light target nuclei are detected at forward angles and are used both as a clean trigger for selecting γ-rays and to normalize to the integrated beam current through Rutherford scattering. Due to the relatively high velocity of the reaction products, especially in inverse kinematics reactions, the γ-rays will be emitted from fast moving nuclei (Figure 6). Therefore, the energy of the γ-ray will depend on the observation angle with respect to the velocity vector of the emitting nucleus following the Doppler Effect equations.

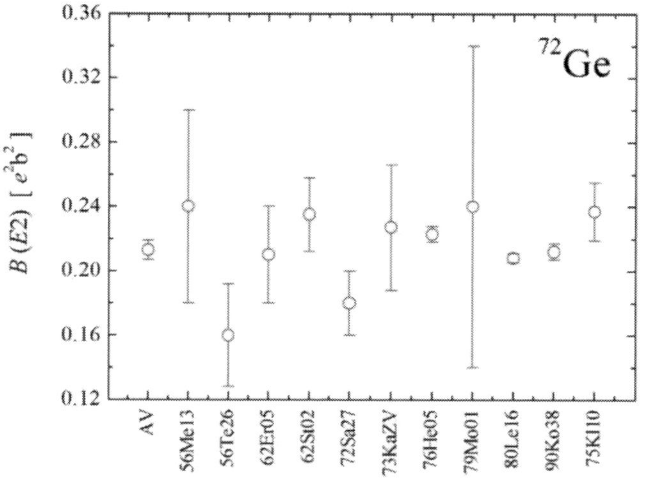

Figure 5. Compilation of B(E2) values for ^{72}Ge (N = 40) from Reference[18].

A very successful research program at the HRIBF ISOL facility is the Coulomb excitation study of n-rich Ge and Se isotopes [2,19]. These results, together with those from CERN's ISOLDE on Zn [20] are providing direct information on the shell closure at N = 50 [21]. In the mass A = 130 region we have measured the B(E2) for even-even isotopes of $^{126\text{-}134}$Sn and $^{132\text{-}136}$Te. Highlights include the first Coulomb excitation measurement of ^{132}Sn which shows the strongest shell closure among the magic nuclei. The results revealed a surprisingly large B(E2) value, further supporting the close similarity between this nucleus and its heavier analog, ^{208}Pb [22].

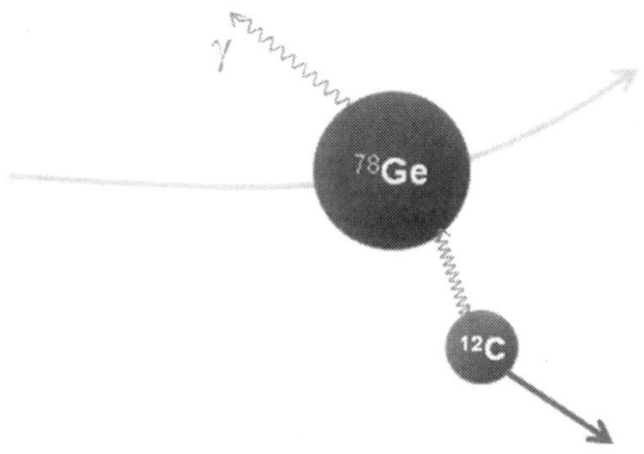

Figure 6. Schematic of the Coulomb excitation in inverse kinematics of the radioactive ^{78}Ge nuclei bombarding a carbon target. The gamma-rays are detected using the HPGe detectors from the CLARION array and the carbon recoils are identified using the CsI(Tl) detectors of the BAREBALL array.

Journal of Physics: Conference Series **387** (2012) 012013 doi:10.1088/1742-6596/387/1/012013

Pioneer experiments were done to measure g-factors of exotic nuclei using the technique of recoil-in-vacuum (RIV) [23]. A transient field g-factor measurement on the first excited state in ^{132}Te using Coulomb excitation has been carried out [24]. Very recently Coulomb excitation experiments where done with ^{124}Sn (stable) and ^{126,128}Sn (unstable) to obtain high precision B(E2) values and the first determination of a static quadrupole moment for the 2^+ state of unstable Sn isotopes [25]. Extension of these measurements to lighter Sn isotopes is underway and would be very valuable for a better understanding of nuclear structure along the Z = 50 shell closure between ^{100}Sn and ^{134}Sn (Figure 8).

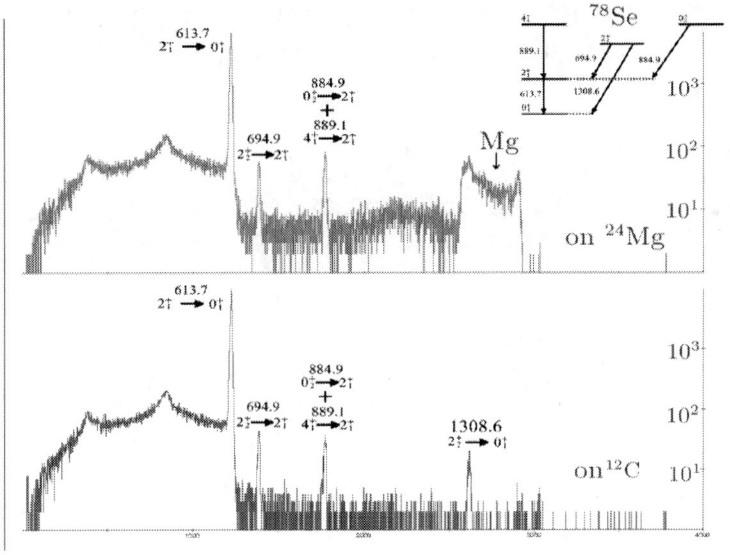

Figure 7. Coulomb excitation spectra for (unstable) ^{78}Ge. Measurements were taken alternating ^{12}C and ^{24}Mg targets during the duration of the run to obtain the static quadrupole moment.

Similarly we hope that the detailed spectroscopic study of the evolution of the N = 50 shell closure will soon be within reach of experiments for the entire isotonic chain from the very neutron-rich ^{78}Ni (Z = 28) to the self-conjugate ^{100}Sn (Z = 50). Our studies of ^{82}Ge and more recently ^{84}Se together with the ^{80}Zn studied at ISOLDE plus revisited measurements on the available N=50 stable isotones allow an excellent opportunity to investigate the role of the proton $f_{5/2}$, $p_{3/2}$, $p_{1/2}$, and $g_{1/2}$ orbitals in the evolution of the N=50 shell.

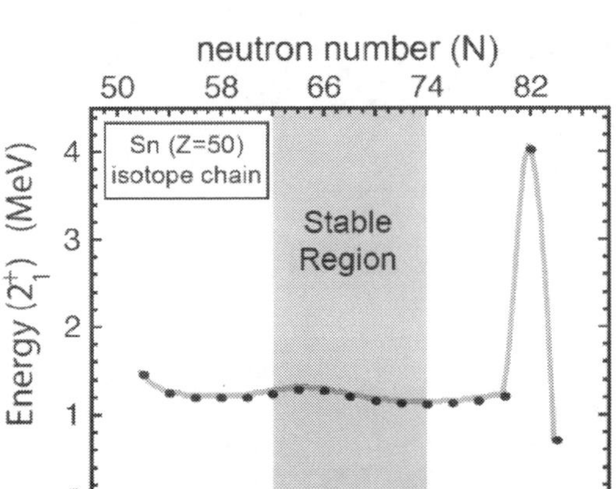

Figure 8. Energies of the 2^+ levels in the even-even Sn isotopes. Spectroscopic studies will be possible in the full range. We have measured the B(E2) values for the neutron-rich Sn isotopes up to ^{134}Sn.

Recent experiments at HRIBF include, Coulomb excitation of *n*-rich nuclei along the N = 50 shell closure and Z = 50 shell closure, the static quadrupole moment of the first 2^+ in ^{78}Ge and $^{126\text{-}128}$Sn and g-factor measurements of a few n-rich isotopes. It is expected that precise measurements of Coulomb excitation using particle gamma-techniques, with both stable and radioactive beams, will provide crucial tests of theoretical predictions, such as the evolution of shell structures in neutron-rich nuclei.

Acknowledgments

I would like to thank my colleagues J.M. Allmond, E. Padilla-Rodal and R. Garcia for useful discussions. Research sponsored by the Office of Nuclear Physics, U.S. Department of Energy and by CONACyT-103366.

References

[1] Radford D C *et al.* 2002 *Phys Rev Lett* **88** 222501
[2] Padilla-Rodal E *et al.* 2005 *Phys Rev Lett* **94** 122501
[3] Garcia R *et al.* 2012 private communication
[4] Radford D C *et al.* 2002 *Eur Phys J* A **15** 171
[5] Galindo-Uribarri A and Urrego-Blanco J P 2007 *Rev Mex Fis* **S53** 35
[6] Beene J R *et al.* 2011 *J Phys* G **38** 024002
[7] Galindo-Uribarri A 1992 *Prog Part Nucl Phys* **28** 463
[8] Galindo-Uribarri A *et al.* 1998 *Phys Lett* B **422** 45
[9] Svensson C E *et al.* 1997 *Nucl Instrum Meth* A **396** 228
[10] Cameron J A *et al.* 1996 *Phys Lett* B 387 266
[11] Caurier E *et al.* 1994 *Phys Rev* C **50** 225
[12] Caurier E *et al.* 1995 *Phys Rev Lett* **75** 2466
[13] Svensson C E *et al.* 2000 *Phys Rev Lett* **85** 2693; Svensson C E *et al.* 2001 *Phys Rev* C **63** 061301R
[14] Ideguchi E *et al.* 2001 *Phys Rev Lett* **87** 222501
[15] de Boer J and Eichler J 1968 *Advances in Nucl Phys* edited by Baranger M and Vogt E (Plenum Press, Inc., New York) vol 1 pp 1
[16] Bass R 1980 *Nuclear Reactions with Heavy Ions* (Springer Verlag, Berlin)
[17] Czosnyka T, Cline D, Wu C Y 1980 *Reference Manual for Coulomb Excitation Data Analysis Code GOSIA* (University of Rochester)
[18] Raman S *et al.* 2001 *At Data Nucl Data Tables* **78** 1; http://www.nndc.bnl.gov/be2/
[19] Galindo-Uribarri A *et al.* 2009 *AIP Conf Proc* **1099** 824
[20] Van de Walle J 2007 *Phys Rev Lett* **99** 142501
[21] Aystö J 2008 *Nucl Phys* A **805** 162c
[22] Varner R L *et al.* 2005 *Eur Phys J* A **25** 391
[23] Stone N J et al. 2005 Phys Rev Lett **94** 192501
[24] Benczer-Koller N *et al.* 2008 *Phys Lett* B **664** 241
[25] Allmond J M *et al.* 2011 *Phys Rev* C **84** 061303R

Gamma-ray spectroscopy at TRIUMF-ISAC: recent highlights and future plans

Gordon C. Ball *for the 8π/TIGRESS collaboration*[1]

TRIUMF, 4004 Wesbrook Mall, Vancouver, B. C. V6T2A3, Canada

E-mail: ball@triumf.ca

Abstract. High-resolution gamma-ray spectroscopy is essential to fully exploit the unique scientific opportunities at the next generation radioactive beam facilities such as TRIUMF Isotope Separator and Accelerator (ISAC). At ISAC the 8π spectrometer and its associated auxiliary detectors is optimized for β-decay studies while TIGRESS, an array of segmented clover HPGe detectors has been designed for reaction studies with accelerated radioactive ion beams. This paper gives a brief overview of these facilities, recent results from the diverse program of nuclear structure and fundamental interaction studies they support and future plans.

1. Introduction

The availability of a wide variety of intense beams of exotic nuclei from the next generation of radioactive ion beam facilities such as the Isotope Separator and Accelerator (ISAC) facility at TRIUMF provides an unprecedented opportunity to address key questions of current interest in nuclear astrophysics, nuclear structure physics and fundamental symmetries. Gamma-ray spectroscopy is a powerful and versatile tool that is essential to all three areas of research at ISAC.

Short-lived isotopes are produced at ISAC by the ISOL (on-line isotope separation) method using a beam of up to 100uA of 500 MeV protons from the TRIUMF H^- cyclotron to bombard thick production targets [1]. The targets can be coupled to a wide variety of ion sources including: surface, laser (TRILIS) [2] and plasma (FEBIAD) sources, to produce the world's most intense RIB beams for certain isotopes such as ^{11}Li. The first UC_x production target was run in December 2010. This target produced high yields of short-lived neutron-rich and actinide isotopes. A license upgrade to operate UC_x targets at beam currents up to 10 μA for a total of 5000μA hr was obtained in Dec 2012.

In 2001, an RFQ and variable energy DTL provide reaccelerated radioactive beams at energies from 0.15-1.8A MeV for nuclear reaction studies of importance in explosive nucleosynthesis environments such as Novae and X-ray bursts. Since January 2007 a Superconducting LINAC installed at ISAC-II [3] has made nuclear reaction studies possible with radioactive beams at energies up to 5 A MeV for $A < 30$. The potential for nuclear structure studies at ISAC-II was greatly enhanced in the summer of 2010 with the installation of high-beta cavities [4] which when combined with a charge state booster will provide radioactive beams up to 7A MeV for $A < 150$.

At ISAC we operate two gamma-ray spectrometers: 1) the 8π gamma ray spectrometer which has been optimized to study the β-decay of low energy (< 60 keV) radioactive beams and TIGRESS

[1] The 8π/TIGRESS international collaboration is a consortium of researchers from 43 institutions in Canada, USA, Europe, Asia, Australia and Africa

(TRIUMF-ISAC Gamma-Ray Escape Suppressed Spectrometer) a next generation array of segmented clover HPGe detectors that is specifically designed to meet the challenges of experiments with high-energy radioactive ion beams. An overview of these facilities, recent highlight from the research programs they support and future plans are presented.

2. Beta-decay studies with the 8π spectrometer

Over the past decade the 8π gamma-ray spectrometer has been dedicated to beta-decay studies with stopped radioactive beams at ISAC-I. The 8π is a close-packed array of 20 Compton-suppressed HPGe detectors each with a relative efficiency of ~25%, resulting in a total photopeak efficiency of 1% at 1.33 MeV. The sensitivity of the spectrometer was improved significantly by the addition of several auxiliary detectors. The β particles are detected with ~80% efficiency using SCEPTAR (Scintillating Electron Positron Tagging Array), a compact array of twenty 1.6 mm thick plastic scintillator detectors arranged in four pentagonal rings centered at approximately the same angles with respect to the beam as the four rings of HPGe detectors. This geometry provides for both high β-γ detection efficiency and bremsstrahlung background suppression, by vetoing those events where the β particle and the γ-ray are detected in the corresponding plastic and HPGe detector pairs.

The SCEPTAR array is mounted inside a spherical Delrin vacuum chamber divided into two hemispheres for easy access. An integral part of SCEPTAR is a computer controlled moving tape collector system. The low-energy (typically ~ 30 keV) beams from ISAC are focused at the centre of the SCEPTAR chamber and deposited onto a 12.7 mm wide collector tape that is fed from a large aluminum storage chamber connected to the vacuum chamber containing the downstream half of SCEPTAR. All aspects of the counting cycle can be controlled including beam on time, measurement time, tape movement and dwell time. A 5 cm thick lead shielding wall located immediately in front of the storage chamber shields the HPGe detectors from long-lived activity remaining on the tape.

The upstream half of SCEPTAR can be replaced by a pentagonal array of five SiLi detectors (PACES) for conversion electron spectroscopy measurements. In addition, a 10 element BaF_2 /$LaBr_3$ array (DANTE) has been developed to provide fast timing for lifetime measurements.

The data obtained from the HPGe, SCEPTAR, PACES and DANTE arrays are collected by independent FERA data streams and software selectable triggers allow for a complete range of singles and coincidence operating modes. A more complete description of the 8π spectrometer facility can be found elsewhere [5,6]

2.1 Superallowed β-decay studies

The precision measurements of the ft values for superallowed $0^+ \rightarrow 0^+$ Fermi β-decays provide demanding tests of the Standard Model description of electroweak interactions. Such measurements have, for example, confirmed the conserved vector current (CVC) hypothesis at the level of 1.3×10^{-4}, set the most stringent limit on fundamental weak scalar interactions at (0.11 ± 0.13)% of the vector strength, and, together with the Fermi coupling constant G_F from muon decay, provide the most precise determination of the Vud element of the Cabibbo-Kobayashi-Maskawa (CKM) quark-mixing matrix [7-9]. A significant uncertainty in these fundamental tests, however, arises from the theoretical nuclear structure corrections that are required to account for the breaking of isospin symmetry by Coulomb and charge-dependent nuclear forces [9-12]. A re-evaluation of these corrections in 2008 [12] incorporating additional physics of the core orbitals led to significant changes, which, when compared to their previously adopted values [11], lower the world-average corrected superallowed Ft value by more than 3σ and raise Vud by 1.5σ; the largest changes in these quantities in more than 20 years. These changes have led to new studies of the isospin symmetry breaking corrections in superallowed decays by a wide variety of theoretical approaches [13-17]. In order to test these models of isospin symmetry breaking, and have confidence in the CVC test and the extraction of Vud that depend on them, high-precision experimental superallowed decay data are now more important than ever.

Journal of Physics: Conference Series **387** (2012) 012014 doi:10.1088/1742-6596/387/1/012014

The 8π spectrometer provides a unique tool for superallowed β-decay branching ratio measurements as the Compton-suppressed HPGe array is sensitive to γ-ray transitions originating from extremely weak β-decay branches (of order 10^{-6}), while the high-efficiency of the SCEPTAR array provides a simultaneous measurement of the total β activity.

In the case of the A≥ 62 superallowed decays, a large number of excited 1^+ states (100 for ^{62}Ga) are predicted to lie within the Q-value window, all of which can be fed via extremely weak beta decay branches, that could not possibly all be observed experimentally. While missing any one of these weak branches would contribute a negligible bias, the sum of all of these missed branches would represent a considerable systematic loss of total decay intensity in a process described as the "Pandemonium effect". One can instead use the low-lying excited 2^+ states in the daughter as collectors for the gamma-decay flux from the weak and unobserved beta-decay branches to the high-lying 1^+ states. The power of this technique with the unique 8π experimental facility was demonstrated previous for the superallowed β-emitter ^{62}Ga where the total superallowed branching ratio was determined to be BR = 99.858(8) %. This measurement was an order of magnitude more precise than the previous adopted value and was pivotal in establishing the ^{62}Ga *ft* value to the level of precision that rivals the best measured cases [18,19].

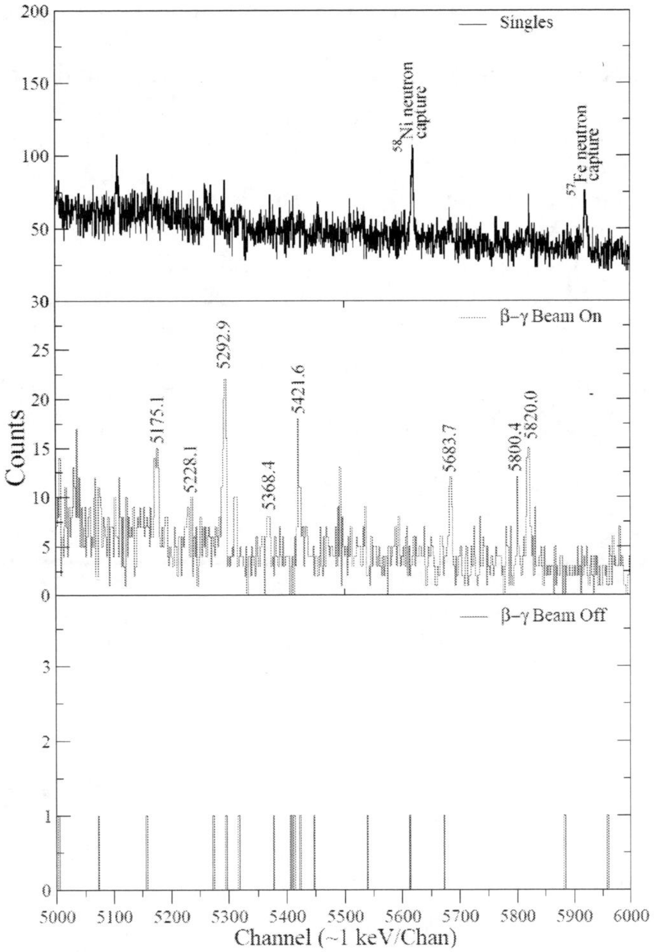

Figure 1. The ^{74}Kr γ -ray spectra between 5000 and 6000 keV. The top panel is a singles γ-ray spectrum while the bottom two panels are bremsstrahlung suppressed β-γ coincidence spectra.

The branching ratio for the superallowed β-emitter ^{74}Rb was determined previous by a similar technique using two plastic scintillators to detect β-particles, two Si(Li) detectors to detect conversion electrons from the decay of the first excited 0^+ level in ^{74}Rb and a single 80% HPGe detector. The value obtained was BR = 99.5(1) % [20]. A new measurement of the branching ratio for ^{74}Rb was carried out in Nov 2010 using the 8π spectrometer. A 100 μA beam of 500MeV protons impinging on a Nb foil target was used to produce a ^{74}Rb beam of 10000 ions/s. A total of 8 x 10^8 positrons were detected by SCEPTAR. A high energy (5000-6000 keV) portion of the bremsstrahlung suppressed β-γ coincidence spectrum is shown in Figure 1. The gamma rays observed at 5684, 5228 and 5175 keV correspond to the decay of a 1^+ level in ^{74}Kr at 5684 keV decaying to the ground, first excited 2^+ and first excited 0^+ levels in ^{74}Kr with intensities of 18,13 and 15 ppm, respectively. A total of 47 new gamma-ray transitions and 15 new levels (i.e nine 1^+, five 2^+ and two 0^+) in ^{74}Kr were observed to follow the β-decay of ^{74}Rb which provided a measured non-analog intensity feeding the ^{74}Rb ground state of $I^{obs}gs$= 0.3958(70) %. When combined with a shell-model calculation to provide a theoretical estimate for the total unobserved ground state intensity this measurement should result in a factor of 3 reduction in the uncertainty of the total superallowed branching ratio for ^{74}Rb [21]. Presently the uncertainty in the experimental *ft* value for ^{74}Rb is dominated by the precision of the measured ^{74}Rb mass. This is expected to be reduced substantially once TITAN completes a high-charge state mass measurement for this nucleus [22].

The *ft* values of the low-Z superallowed β-emitters, i.e. ^{10}C and ^{14}O, are of particular interest since they are ones most sensitive to a possible scalar current interaction [8]. The half-life of ^{14}O has been determined with a precision of ± 0.02% from the weighted average of 8 measurements, two by direct β-counting and the remaining six by counting the 2.3 MeV γ ray photopeaks in the daughter nucleus ^{14}N, that following 99.4% of the β-decays in ^{14}O. However, a closer examination of these measurements reveals that the weighted average values of the half-life of ^{14}O determined by the β and γ counting methods differ by 2.6 σ. A high priority program led by G.F Grinyer [23] to improve limits on weak scalar interactions through precise β and γ-ray based lifetime measurements for the superallowed emitter ^{14}O, exploits the new technique developed for the 8π spectrometer to account for detector pulse pile-up effects in high-precision β-decay lifetime measurements via γ counting [24]. The β-counting of ^{14}O was measured simultaneously using a zero-degree scintillator that replaced the downstream half of SCEPTAR. Representative decay curves obtained by β and γ counting in a test "proof of principle" run carried out in November 2011 are shown in Figures 2 and 3. A longer run to achieve the required statistical precision is planned for 2012.

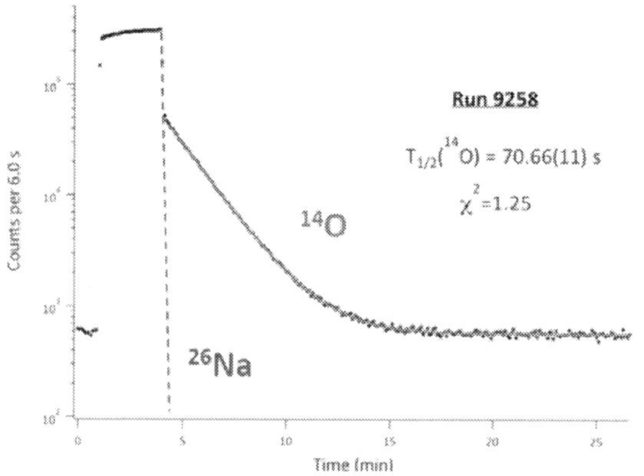

Figure 2. Typical decay curve obtained by β-counting using the zero degree scintillator.

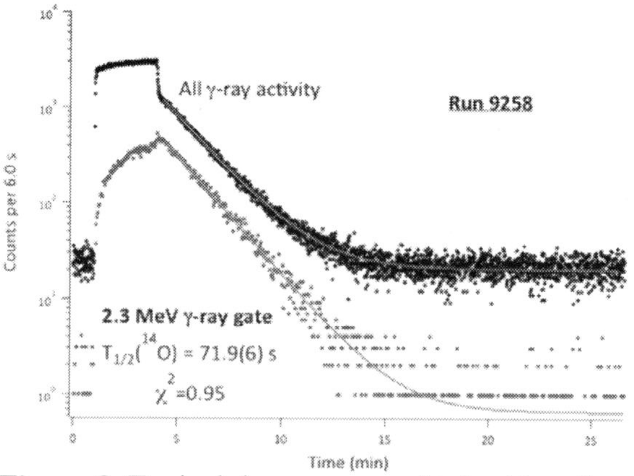

Figure 3. Typical decay curves obtained by all γ-ray activity and by gating on the 2.31 MeV γ-ray photopeak from the daughter nucleus ^{14}N.

2.2 Nuclear structure studies

A diverse program of nuclear structure studies have been carried out over the past decade using the 8π spectrometer including: 1) the study of coexisting collective phases in N = 90 nuclei, ^{156}Dy, ^{158}Er, and ^{160}Yb, populated in the β-decay of ^{156}Ho, ^{158}Tm and ^{160}Lu, respectively [25], 2) a systematic study of the decay of ^{110,112}Ag that provided a rigid test of the validity of the U(5) description (modified by intruder-phonon mixing) in ^{110,112}Cd [26] and the β-decay of the neutron-rich nuclei ^{32}Na[27] and ^{11}Li [28].

The production of high-intensity neutron-rich and actinide beams from UC_x production targets has led recently several new initiatives. In particular, a detailed study of multiphonon excitations and mixed-symmetry states in ^{94}Zr, was carried out via the β-decay of ^{94}Y. In this experiment a beam of $\sim 10^8$ ^{94}Rb ions/s (t1/2 = 2.3s) was implanted onto the tape at the center of the 8π spectrometer array for about 30 minutes to produce a ~5 x10^5/s source of the granddaughter nucleus ^{94}Y (t1/2 = 18.7 min). This high-statistics γ-γ coincidence data set is presently under analysis [29]. Another experiment focused on the study of neutron-rich Rb isotopes up to ^{102}Rb, on the boundaries of the r-process path [30]. In this case the experiment was operated in a cycling mode where the beam was implanted onto the tape and allowed to decay for a short period of time before moving the tape to remove the longer-lived isobaric nuclides. This method allowed for separation of various isobars in the cocktail beam. This is illustrated in Figure 4 for an A =100 beam where the time profiles of γ-rays can be associated with transitions of a nuclide with the appropriate half-life.

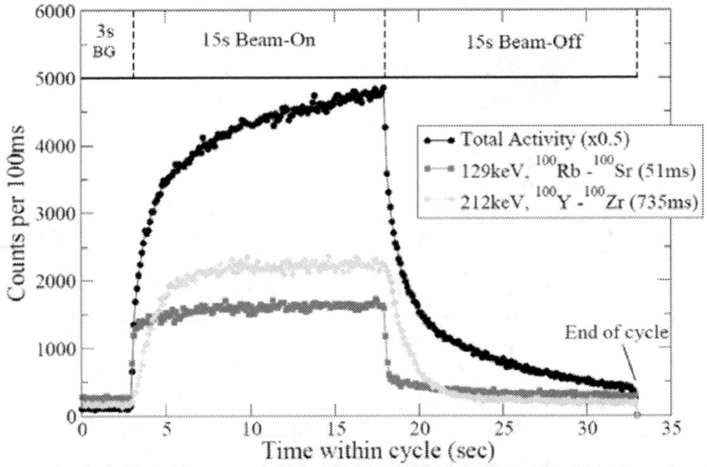

Figure 4. Activity from an A = 100 beam from a UCx target measured as a function of the time within the cycle. Each cycle consists of 3 seconds of background, 15 seconds of beam on and 15 seconds of beam off counting. The assignment of γ-rays to the decay of specific nuclides based on their time profile is illustrated by gates on the $2^+ \rightarrow 0^+$ transitions in ^{100}Sr and ^{100}Zr.

3. GRIFFIN

The recent funding of GRIFFIN (Gamma-Ray Infrastructure For Fundamental Investigations of Nuclei) will dramatically upgrade the decay spectroscopy capabilities at ISAC-I [6,31]. GRIFFIN will consist of an array of sixteen large-volume HPGe clover detectors arranged in a close-packed configuration with a total singles absolute efficiency of ~17% at 1.3 MeV. It has been designed to couple to all existing 8π ancillary detector systems and is also designed to work with the DESCANT neutron detector array (see section 4.1 for details) for β-delayed neutron decay studies. The new digital data acquisition system will operate at large data through-put in a semi-triggerless mode to facilitate both high-rate nuclear structure and precision measurements to the level of better than 0.05%. GRIFFIN is scheduled for early implementation in late 2014.

4. TIGRESS

The gamma-ray spectroscopy program at ISAC-II is centered on the TRIUMF-ISAC Gamma-Ray Escape-Suppressed Spectrometer (TIGRESS), a next generation gamma-ray spectrometer designed for use with accelerated radioactive beams at energies up to 7A MeV for A < 150. TIGRESS is comprised of up to 16 segmented clover-type HPGe detectors, each equipped with a composite BGO-CsI Compton Suppression shield. Sensitivity to the location of gamma-ray interactions within the HPGe clover is achieved through 32-fold segmentation of the outer electrical contacts, and through pulse shape analysis following 100 MHz digitization of the signal waveforms from each contact . This position sensitivity allows the detectors to be placed close to the target location without sacrificing angular resolution, leading to large gains in gamma-ray detection efficiency while retaining the high intrinsic resolution of the HPGe clover detectors. A picture of TIGRESS installed in ISAC-II is shown in Figure 5. More detailed descriptions of TIGRESS can be found elsewhere [32,33].

Figure 5. A picture of TIGRESS installed in the ISAC-II experimental hall.

Figure 6. A schematic view of DESCANT coupled to TIGRESS/GRIFFIN by replacing the downstream lampshade.

4.1 Coulomb excitation studies with Bambino

Light-ion Coulomb excitation studies have been carried out at TIGRESS using a charged particle detector system Bambino, designed to accommodate annular (CD) double-sided silicon detectors both upstream and downstream of the target. In these experiments the B(E2) transition matrix elements for the light-ion beams were determined relative to those for target nuclei with well-known transition strengths. The first experiment carried out during the early implementation phase of TIGRESS was a study of the mirror pair ^{21}Ne and ^{21}Na which resolved a previous discrepancy in ^{21}Na and confirmed that the low-lying states of the mirror pair can be well described within pure *sd* model space by current effective interactions [34]. Other studies included the first Coulomb excitation measurements for ^{20}Na [35] and ^{29}Na [36].

Recently, a reorientation-effect Coulomb-excitation measurement of the sign of the spectroscopic quadrupole moment the first excited 2^+ state at 3.368 MeV in ^{10}Be has been carried out using a beam of 10^7 ^{10}Be ions/s at 41 MeV to bombard a ^{194}Pt target. A preliminary analysis of these data gives a value of $Q_s(2^+) = -0.06 \pm 0.07$ eb in agreement with new no-core shell model calculations with the *CD-Bonn 2000* two-nucleon potential and large shell model spaces. The one neutron halo nucleus ^{11}Be has the strongest known E1 strength between bound states and the energy weighted E1 strength up to 4 MeV exhausts nearly 70% of the cluster sum rule. A low energy Coulomb excitation measurement to determine the E1 strength with high precision and accuracy was carried out using beams of 10^6 ^{11}Be ions/s at 19, 23 and 41 MeV to bombard a ^{196}Pt target[38]. The results of this experiment will be compared with the predictions of no-core shell model plus resonating group method [38].

4.2 Transfer reactions studies with SHARC

The Silicon Highly-segmented Array for Reactions and Coulex (SHARC), a new multipurpose array for charged particle detection has been developed under the leadership of the University of York [39]. The array built for use with TIGRESS was designed to be compatible with transfer and inelastic scattering in inverse kinematics, fusion evaporation, Coulomb excitation and deep-inelastic scattering experiments. The highly segmented silicon detector array consists of two boxes of DSSSDs one on either side of 90 degrees plus upstream and downstream QQQ2 CD-detectors. The first experiment with SHARC was a measurement of the $^{25}Na(d,p)^{26}Na$ reaction as part of a program to follow the evolution of the shell structure of neutron-rich *sd*-shell nuclei [40]. In this study a high intensity beam of 3 x 10^7 ^{26}Na ions at 5A MeV was used to bombard a CD_2 target. The γ-rays emitted following the decay of low-lying levels in ^{26}Na were detected with TIGRESS in coincidence with protons detected with SHARC to facilitate the measurement of angular distributions for states populated in ^{26}Na that couldn't be resolved with SHARC alone.

4.3 Auxiliary detector development for TIGRESS

Several new particle detection systems are under development to exploit the research opportunities with TIGRESS including a CsI(Tl) light charged-particle detector array with an integrated plunger device (TIP) for lifetime measurements to be used in fusion-evaporation studies with TIGRESS [41], SPICE (SPectrometer for Internal Conversion Electrons), a high-efficiency high-energy conversion electron spectrometer, a 70-element DEuterated SCintillator Array for Neutron Tagging (DESCANT) (see Figure 6) [42], and the ElectroMagnetic Mass Analyser (EMMA) for heavy ion recoil separation and detection [43]. All of these detector systems are scheduled for completion within the next two years.

5. Summary and outlook

During the last five years there has been a rapid development in the capabilities of ISAC to provide both low energy and reaccelerated radioactive beams for nuclear physics research. In parallel the construction of TIGRESS, a state of the art gamma-ray spectrometer, together with a suite of auxiliary detectors, and the ongoing development of GRIFFIN to replace the 8π spectrometer, will enable the national and international ISAC user community to remain at the forefront in RIB research for the foreseeable future.

Finally, the potential for nuclear structure studies at ISAC will be greatly enhanced with construction of ARIEL, the Advanced Rare Isotope Laboratory [44]. Phase I of this project scheduled for completion in 2015 is the construction of a 25 MeV 100 kW superconducting RF electron accelerator for the production of neutron rich isotopes from the photo-fission of actinide targets. Phase two of the ARIEL project includes: increasing the e-linac energy to 50 MeV at 500 kW beam power, constructing a second 500 MeV proton beam line, and two target stations and front end mass separators; with the goal of delivering two additional RIBs simultaneously.

Acknowledgements

The author gratefully acknowledges the invaluable contributions made by the TRIUMF accelerator and beam development teams and the ISAC experimental support and operations staff. He would also like to thank members of the 8π/TIGRESS collaboration for their contributions to this overview of the ISAC gamma-ray spectroscopy program. This work was supported by the Natural Sciences and Engineering Research Council of Canada. TRIUMF receives federal funding via a contribution agreement through the National Research Council of Canada.

References

[1] Bricault P *et al.* 2003 *Nucl. Instrum. Methods B* **204** 319
[2] Lassen J *et al.* 2009 *Solid-State Laser, Resonant Ionization Laser Ion Source (RILIS) and Laser Beam Transport at RIB Facilities* (AIP Conference Proceedings 1099(1)) 769
[3] Laxdal R E 2007 *Proceedings of PAC07*, Albuquerque, New Mexico, USA 2593
[4] Laxdal R E *et al.* 2008 *Proc. of LINAC08*,Victoria, British Columbia, Canada) 97
[5] Ball G C *et al.* 2005 *J. Phys. G. Nucl. Part. Phys.* **31** S1491
[6] Garnsworth A *et al.* 2011 *Proc. 14th Intl. Sym. On Capture γRays and Related Topics*, Guelph, Ontario and to be published.
[7] Marciano W J and Sirlin A 2006 *Phys. Rev. Lett.* **96** 032002.
[8] Hardy J C and Towner I S 2009 *Phys. Rev. C* **79** 055502
[9] Towner I S and Hardy J C, 2010 *Rep. Prog. Phys* **73** 046301
[10] Ormand W E and Brown B A 1995 *Phys. Rev. C* **52** 2455
[11] Towner I S and Hardy J C 2002 *Phys. Rev. C* **66** 035501
[12] Towner I S and Hardy J C 2008 *Phys. Rev. C* **77** 025501
[13] Miller G A and Schwenk A 2008 *Phys. Rev. C* **78** 035501; 2009 *Phys. Rev. C* **80** 064319
[14] Calik A E et al. 2009 *Z. Naturforsch.* **64a**, 865
[15] Satula W *et al.* 2009 *Phys. Rev. Lett.* **103** 012502; 2011 *Phys. Rev. Lett.* **106** 132502
[16] Liang H *et al.* 2009 Phys. Rev. C **79** 064316
[17] Towner I S and Hardy J C 2010 *Phys. Rev. C* **82** 065501
[18] Finlay P *et al.* 2008 *Phys Rev C* **78** 025502
[19] Hyland B *et al.* 2006 *Phys. Rev. Lett.* **100** 192504
[20] Piechaczek A *et al.* 2003 *Phys. Rev. C* **67** 051305(R)
[21] Dunlop R *et al.* 2011 *Proc. 14th Intl. Sym. on Capture γ-Rays and Related Topics*, Guelph, Ontario and to be published.
[22] Ettenauer S *et al.* 2011 *Phys. Rev. Lett.* **107** 272501
[23] Grinyer G *et al.* to be published
[24] Grinyer G *et al.* 2007 *Nucl. Inst. Meth. A* **579** 1005
[25] Kulp D et al. to be published
[26] Garrett P E et al. 2011 *Acta Physica Polonica B* **42** 799 and to be published
[27] Matton C M *et al.* 2007 *Phys. Rev. C* **75** 017302
[28] Matton C M *et al.* 2009 *Phys. Rev. C* **80** 034318
[29] Yates S *et al.* to be published
[30] Garnsworthy A *et al.* to be published
[31] Svensson C E and Garnsworthy A to be published
[32] Svensson C E 2005 *et al.* J. Phys. G **31** S1663
[33] Ball G C *et al.* 2011 J. Phys. G **38** 024003 and references cited therin
[34] Schumaker M A *et al.* 2008 *Phys Rev C* **78** 044321
[35] Schumaker M A *et al.* 2009 *Phys Rev C* **80** 044325
[36] Hurst A M *et al.* 2009 *Phys Lett B* **674** 168.
[37] Orce N and Navratil P *et al* to be published
[38] Kwan E *et al* to be published and Quaglioni S and Navratil P 2008 *Phys. Rev. Lett.* **101** 092501
[39] Diget C A *et al.* 2011 *J. Inst. G* **6** P02005
[40] Wilson G L 2011 *Proc. of the Rutherford Centenial Conf. on Nucl. Phys.* Manchester UK
[41] Starosta K et al to be published
[42] Garrett P E to be published
[43] Davids B and Davids C N 2005 *Nucl. Inst. Meth. A* **544** 565
[44] Merminga L *et al.* 2011 *Proc. of IPAC*, San Sebastian, Spain **WEOBA01** 1917

Journal of Physics: Conference Series **387** (2012) 012015 doi:10.1088/1742-6596/387/1/012015

Density functional approaches to nuclear dynamics

T Nakatsukasa[1,2], S Ebata[1,3], P Avogadro[1,4], L Guo[1,5], T Inakura[1] and K Yoshida[6]

[1] RIKEN Nishina Center, 2-1 Hirosawa, Wako-shi, 351-0198, Japan
[2] Center for Computational Sciences, University of Tsukuba, Tsukuba 305-8571, Japan
[3] Center for Nuclear Study, University of Tokyo, Wako-shi 351-0198, Japan
[4] Department of Physics & Astronomy, Texas A&M University-Commerce, Commerce, Texas 75428, USA
[5] College of Physical Sciences, Graduate University of Chinese Academy of Sciences, Beijing 100049, China
[6] Department of Physics, Niigata University, Niigata 950-2181, Japan

Abstract. We present background concepts of the nuclear density functional theory (DFT) and applications of the time-dependent DFT with the Skyrme energy functional for nuclear response functions. Practical methods for numerical applications of the time-dependent Hartree-Fock-Bogoliubov theory (TDHFB) are proposed; finite amplitude method and canonical-basis TDHFB. These approaches are briefly reviewed and some numerical applications are shown to demonstrate their feasibility.

1. Introduction

The nucleus is a quantum object. The nuclear interaction is not strong enough to localize the nucleonic wave function, thus, the nuclear matter stays in the liquid phase even at zero temperature [1]. This strong quantum nature in nuclei leads to a rich variety of unique phenomena. Extensive studies have been made for constructing theoretical models to elucidate basic nuclear dynamics behind a variety of nuclear phenomena. Simultaneously, significant efforts have been made in the microscopic foundation of those models.

Although there have been significant developments in the "first-principles" large-scale computation, starting from the bare nucleon-nucleon (two-body & three-body) forces, they are still limited to light nuclei with small mass numbers, typically $A \lesssim 10$ [2]. In contrast, the density functional theory (DFT) is currently a leading theory for describing nuclear properties of heavy nuclei [3, 4]. It is capable of describing almost all nuclei, including nuclear matter, with a single universal energy density functional (EDF). In addition, its strict theoretical foundation is given by the basic theorem of the DFT [5, 6]. Since the nucleus is a self-bound system without an external potential, the DFT theorem should be modified from its original form. This problem was addressed by recent studies [7, 8, 9]. In this paper, we present basic concepts of the EDF in nuclei and feasible methodologies of the time-dependent DFT.

2. Backgrounds of nuclear energy density functionals

In this section, basic properties of nuclei and historical developments in nuclear structure theory, which leads to the nuclear energy density functional, are briefly reviewed. To simplify the discussion here, we consider an infinite uniform nuclear matter neglecting the Coulomb interaction.

2.1. Basic property of nuclear systems: Saturation and independent-particle motion

2.1.1. Saturation The volume and total binding energy of observed nuclei in nature are approximately proportional to the mass number A. In other words, they have an approximately constant density $\rho_0 \approx 0.17$ fm^{-3} and a constant binding energy per particle $B/A \approx 8$ MeV. Thus, extrapolating this property to the infinite nuclear matter neglecting the Coulomb interaction, the nuclear matter should have an equilibrium state with $\rho_0 \approx 0.17$ fm^{-3} and $B/A \approx 16$ MeV, at zero pressure and zero temperature. This property are called "saturation property", that is analogous to the liquid. The most famous and successful model based on this liquid picture of nuclei is the empirical mass formula of Bethe and Weizsäcker [10, 11]. This formula contains the surface and Coulomb terms in addition to the leading term proportional to A, which well accounts of the bulk part of the nuclear binding.

2.1.2. Independent-particle motion in nuclei There are many evidences for the fact that the mean-free path of nucleons is larger than the size of nucleus. The great success of the nuclear shell model [12] gives such an example, in which nucleons are assumed to move independently inside an average one-body potential. The scattering experiments with incident neutrons and protons provide more quantified information on the mean-free path. In fact, the mean free path depends on the nucleon's energy, and becomes larger for lower energy [1]. Therefore, it is natural to assume that the nucleus can be primarily approximated by the independent-particle model with an average one-body potential. For the nuclear matter, this approximation leads to the degenerate Fermi gas of the same number of protons and neutrons ($Z = N = A/2$). The observed saturation density of $\rho_0 \approx 0.17$ fm^{-3} gives the Fermi momentum, $k_F \approx 1.36$ fm^{-1}, which corresponds to the Fermi energy (the maximum kinetic energy), $T_F = k_F^2/2M \approx 40$ MeV.

2.2. Problems of a mean-field picture

Evidences of the independent-particle motion encourage us to adopt the mean-field picture of nuclei. However, it turned out that the mean-field models cannot describe the nuclear saturation property. Let us explain this for the uniform nuclear matter with a constant attractive potential $V < 0$.

The constancy of B/A means that it is equal to the separation energy of nucleons, S. In the Fermi-gas model, it is estimated as

$$S \approx B/A \approx -(T_F + V). \tag{1}$$

Since the binding energy is $B/A \approx 16$ MeV, the potential V is about -55 MeV. It should be noted that the relatively small separation energy is the consequence of the significant cancellation between kinetic and potential energies. In the mean-field theory, the total (binding) energy is given by

$$-B = \sum_{i=1}^{A} \left(T_i + \frac{V}{2} \right) = A \left(\frac{3}{5} T_F + \frac{V}{2} \right), \tag{2}$$

where we assume that the average potential results from a two-body interaction. The two kinds of expressions for B/A, Eqs. (1) and (2), lead to $T_F \approx -5V/4 \approx 70$ MeV, which is different from the previously estimated value (~ 40 MeV). Moreover, the negative separation energy ($T_F + V > 0$) contradicts the fact that the nucleus is bound!

To reconcile the independent-particle motion with the saturation property of the nucleus, the nuclear average potential must be state dependent. Allowing the potential V_i depend on

Journal of Physics: Conference Series **387** (2012) 012015 doi:10.1088/1742-6596/387/1/012015

the state i, the potential V should be replaced by that for the highest occupied orbital V_F in Eq. (1), and by its average value $\langle V \rangle$ in the right-hand side of Eq. (2). Then, we obtain the following relation:

$$V_F \approx \langle V \rangle + T_F/5 + B/A. \tag{3}$$

Therefore, the potential V_F is shallower than its average value.

Weisskopf suggested the momentum-dependent potential V, which can be expressed in terms of an effective mass m^* [13]:

$$V_i = U_0 + U_1 \frac{k_i^2}{k_F^2}. \tag{4}$$

Actually, if the mean-field potential is non-local, it can be expressed by the momentum dependence. Equation (4) leads to the effective mass, $m^*/m = (1 + U_1/T_F)^{-1}$. Using Eqs. (1), (3), and (4), we obtain the effective mass given by

$$\frac{m^*}{m} = \left\{ \frac{3}{2} + \frac{5}{2} \frac{B}{A} \frac{1}{T_F} \right\}^{-1} \approx 0.4. \tag{5}$$

Quantitatively, this value disagrees with the experimental data. The empirical values of the effective mass vary according to the energy of nucleons, $0.7 \lesssim m^*/m \lesssim 1$, however, they are almost twice larger than the value in Eq. (5). As far as we use a normal two-body interaction, this discrepancy should be present in the mean-field calculation with any interaction, because Eq. (5) is valid in general for a saturated self-bound system. Therefore, the naive mean-field models have a fundamental difficulty to describe the nuclear saturation.

2.3. Nucleon-nucleon interaction (nuclear force)

To understand the origin of the problem, properties of the nuclear force provides an important key. The saturation property of nuclear density reflects a balance between attractive and repulsive contributions to nuclear binding energy. One source of such repulsive effects is the nucleonic kinetic energy of the Fermi gas. However, its contribution per particle is proportional to $\rho^{2/3}$, which is not strong enough to resist against the collapse caused by the attractive force between nucleons. Therefore, the nucleonic interaction must contain a repulsive element. Indeed, the phase-shift analysis on the nucleon-nucleon scattering at high energy ($E_{\rm lab} > 250$ MeV) reveals a short-range strong repulsive core in the nucleonic force. The radius of the repulsive core is approximately $c \approx 0.5$ fm. This strong repulsive core prevents the nucleons approaching closer than the distance c, which produces a strong two-body correlation, $\rho^{(2)}(\vec{r}_1, \vec{r}_2) \approx 0$ for $|\vec{r}_1 - \vec{r}_2| < c$. The attractive part of the interaction has a longer range, which can be characterized by the pion's Compton wave length λ_π, and is significantly weaker than the repulsion. Thus, a naive application of the mean-field calculation fails to bind the nucleus, since the mean-field approximation cannot take account of such strong two-body correlations.

At first sight, this seems inconsistent with the experimental observations. As we mentioned in Sec 2.1, there are many experimental evidences for the independent-particle motion in nuclei. We may intuitively understand that it is due to the fact that the nucleonic density is significantly smaller than $1/c^3$. Therefore, the collisions by the repulsive core rarely occur and the system can be approximately described in terms of the independent-particle motion. Furthermore, the effects of the Pauli principle hinder the collisions, since the nucleons cannot be scattered into occupied states. Although the repulsive-core collisions are experienced by only a small fraction of nucleons ($\sim \rho_0 c^3$), each collision carries a large amount of energy. Therefore, the repulsive core provides an important contribution to the total energy and are responsible for the saturation.

Another important factor for the independent particle motion is the strong quantum nature due to the weakness of the attractive part of the nuclear force. The importance of the quantum nature can be measured by the magnitude of the zero-point kinetic energy compared to that of the interaction. If the attractive part of the nuclear force were much stronger than the unit of $\hbar^2/Mc^2$, the quantum effect would disappear and each nucleon would stay at the bottom of the interaction potential. Then, the nucleus would crystallize at low temperature. In reality, the attraction of the nuclear force is so weak that it barely produces many-nucleon bound states at the relatively low density.

2.4. From Brueckner theory to EDF

The nuclear matter theory pioneered by Brueckner gives a hint for a solution for the previous difficulty to understand the nuclear saturation. The Brueckner theory may provide a first step toward the quantitative treatment to understand the saturation property and the independent-particle motion in nuclei. Details of the theory can be found at Refs. [14, 15].

The basic ingredient of the Brueckner theory is a two-body scattering matrix of particle 1 and 2 inside nucleus caused by the nuclear force v,

$$G(\omega) \equiv v + v\frac{Q}{\omega - Q(T_1 + T_2)Q}G(\omega), \tag{6}$$

where T_i is the kinetic energy of particle i, Q is the Pauli-exclusion operator to restrict the intermediate states, and ω is called a starting energy that depends on energies of particle 1 and 2. This is called G-matrix [16]. The G-matrix renormalizes high-momentum components in the bare nuclear force and becomes an effective interaction in nuclei under the independent-pair approximation. The G-matrix reflects an underlying structure of the independent many-nucleon system through the operator Q and the starting energy ω. Inevitably, the G-matrix becomes state (structure) dependent.

Since the short-range singularity is renormalized in the G-matrix, we can calculate the total energy in the independent-particle (mean-field) model, analogous to Eq. (2).

$$-B = \sum_{i=1}^{A}\left\{T_i + \frac{1}{2}\sum_{j=1}^{A}\bar{G}_{ij,ij}(\omega_{ij})\right\} \tag{7}$$

where $\omega_{ij} = \epsilon_i + \epsilon_j$, defines the self-consistency condition for the Brueckner's single-particle energies, and $\bar{G}_{ij,ij} \equiv G_{ij,ij} - G_{ij,ji}$. This is called Brueckner-Hartree-Fock (BHF) theory. The validity of the BHF theory is measured by the wound integral $\kappa = \langle\psi - \phi|\psi - \phi\rangle$, where $|\phi\rangle$ is an unperturbed two-particle wave function and $|\psi\rangle$ is a correlated two-particle wave function in nucleus. κ is known to be about 15 %. The BHF calculation was successful to describe the nuclear saturation, however, could not simultaneously reproduce empirical values of B/A and ρ_0, known as a problem of the Coester band [17]. Its applications to finite nuclei also have similar problems to reproduce the energy, radius, and density in the ground state.

A possible solution to these problems was given by Negele [18]. Starting from a realistic G-matrix, using the expressions for the Pauli operator

$$\langle\vec{r}_1\vec{r}_2|Q|\vec{r}_1'\vec{r}_2'\rangle = \left\{\delta(\vec{r}_1 - \vec{r}_1') - \rho(\vec{r}_1 - \vec{r}_1')\right\}\left\{\delta(\vec{r}_2 - \vec{r}_2') - \rho(\vec{r}_2 - \vec{r}_2')\right\}, \tag{8}$$

and the average single-particle energy $\epsilon[\rho(\vec{r})]$, the local density approximation is introduced to expand the off-diagonal density matrix $\rho(\vec{r} + \vec{s}/2, \vec{r} - \vec{s}/2)$ with respect to the relative coordinate $|\vec{s}|$. Then, a short-range part of the G-matrix, which is not fully understood, is phenomenologically added to the energy expression to quantitatively fit the saturation property, and finally, the total energy is treated variationally. This procedure is called the density matrix expansion (DME) [19]. The state dependence of the G-matrix is now replaced by the density dependence. The final result for the total energy, for a uniform nuclear matter, is written as a function of the neutron and proton densities, ρ_n and ρ_p, and the kinetic densities, τ_n and τ_p; $E = E[\rho_n, \rho_p, \tau_n, \tau_p]$. This expression can be generalized for finite nuclei as

$$E = \int d\vec{r} H(\vec{r}), \quad H(\vec{r}) = H[\rho_n(\vec{r}), \rho_p(\vec{r}), \tau_n(\vec{r}), \tau_p(\vec{r})], \tag{9}$$

$H(\vec{r})$ here is a complete analogue of the Hamiltonian density in the Skyrme EDF (without the spin-orbit and Coulomb terms) [20]. Thus, the DME with a microscopic G-matrix leads to a variational treatment of a simple EDF of the Skyrme type.

The essential aspect of DME is in its non-trivial density dependence and the variational treatment. Now, the expression for the total energy, Eq. (2), should be modified due to these non-trivial density-dependent terms. This resolves the previous issue, and provides a consistent independent-particle description for the nuclear saturation. In nuclear physics, this was often interpreted as the density-dependent effective interaction. In this terminology, the variation of the total energy with respect to the density contains re-arrangement potential, $\partial V_{\text{eff}}[\rho]/\partial\rho$, which comes from the density dependence of the effective force $V_{\text{eff}}[\rho]$. These terms turn out to be crucial to obtain the saturation condition.

In summary, the failure in the mean-field description of nuclei using phenomenological effective interactions can be traced back to the missing state (structure) dependence. The EDF approaches take into account the state dependence in terms of the non-trivial density dependence.

3. Time-dependent Hartree-Fock-Bogoliubov theory

For a quantitative description of heavy nuclei in open-shell configurations, it is necessary to include the pairing correlations. This can be done by a straightforward extension of the Skyrme energy functional simply added by the pairing energy functional.

$$E[\rho, \kappa] = E_{\text{Skyrme}}[\rho] + E_{\text{pair}}[\rho, \kappa], \tag{10}$$

where κ is the pair density. The variation of the energy with the constraint on the average particle number leads to the Hartree-Fock-Bogoliubov (HFB) equation [15, 21]. However, here, let us start from the time-dependent Hartree-Fock-Bogoliubov (TDHFB) equation,

$$i\frac{\partial}{\partial t}\mathcal{R}(t) = [\mathcal{H}(t), \mathcal{R}(t)], \tag{11}$$

where $\mathcal{H}(t)$ is the HFB Hamiltonian usually written in a form [15]

$$\mathcal{H}(t) = \mathcal{H}[\mathcal{R}(t)] \equiv \begin{pmatrix} h[\rho(t), \kappa(t)] & \Delta[\rho(t), \kappa(t)] \\ -\Delta^*[\rho(t), \kappa(t)] & -h^*[\rho(t), \kappa(t)] \end{pmatrix}. \tag{12}$$

The generalized density matrix $R(t)$ is given by

$$\mathcal{R}(t) = \begin{pmatrix} \rho(t) & \kappa(t) \\ -\kappa^*(t) & 1 - \rho^*(t) \end{pmatrix} \tag{13}$$

which is Hermitian and idempotent: $\mathcal{R}^2 = \mathcal{R}$. Thus, the eigenvalues of $\mathcal{R}(t)$ are 0 and 1. The eigenvectors $\Psi_\nu(t) = \begin{pmatrix} U_\nu(t) \\ V_\nu(t) \end{pmatrix}$, which correspond to the eigenvalue 0, are called "unoccupied" quasiparticle (qp) orbitals. For every unoccupied qp orbital, there exits a conjugate "occupied" orbital $\bar{\Psi}_\nu(t) = \begin{pmatrix} V_\nu^*(t) \\ U_\nu^*(t) \end{pmatrix}$, whose eigenvalue is 1.

We define the following matrix of size $2M \times M$ where M is the dimension of the adopted single-particle Hilbert space,

$$\Psi_{\alpha\nu}(t) = \begin{cases} \langle\alpha|U_\nu(t)\rangle & \alpha = 1, \cdots, M \\ \langle\alpha - M|V_\nu(t)\rangle & \alpha = M+1, \cdots, 2M \end{cases} \tag{14}$$

which collectively represents time-dependent unoccupied qp orbitals ($\nu = 1, \cdots, M$). The matrix for the occupied orbitals $\bar{\Psi}(t)$ are defined in the same manner, with $\Psi_\nu(t)$ replaced by $\bar{\Psi}_\nu(t)$. Using these matrices, $\mathcal{R}(t)$ can be written in a simple form as $\mathcal{R}(t) = \bar{\Psi}(t)\bar{\Psi}^\dagger(t) = 1 - \Psi(t)\Psi^\dagger(t)$. The orthonormal property of the qp orbitals is given by $\Psi^\dagger(t)\Psi(t) = \bar{\Psi}^\dagger(t)\bar{\Psi}(t) = 1$. Note that $\mathcal{R}(t)$ can be regarded as the projection operator onto the occupied space. Substituting these expressions into Eq. (11), we obtain the TDHFB equation for the qp orbitals as

$$\mathcal{R}(t)\left\{i\frac{\partial}{\partial t} - \mathcal{H}(t)\right\}\Psi(t) = 0, \qquad \mathcal{Q}(t)\left\{i\frac{\partial}{\partial t} - \mathcal{H}(t)\right\}\bar{\Psi}(t) = 0, \tag{15}$$

where $\mathcal{Q}(t) = 1 - \mathcal{R}(t)$ is the projection operator onto the unoccupied space. Therefore, the occupied-occupied and unoccupied-unoccupied matrix elements of $i\partial/\partial t - \mathcal{H}(t)$ are arbitrary at every instant of time. This is related to the $U(M)$ invariance; $\mathcal{R}(t)$ is invariant with respect to the transformation among the occupied (unoccupied) qp orbitals, $\bar{\Psi}' = \bar{\Psi}U$ ($\Psi' = \Psi U$), where U is an $M \times M$ unitary matrix.

Now, let us derive the static Hartree-Fock-Bogoliubov (HFB) equation as a *quasi-stationary* solution of Eq. (11). This does not correspond to the solution for $\partial\mathcal{R}/\partial t = 0$, because the solution must have a given value of the average particle number; $\mathrm{tr}\rho = N$. When the system is in the superfluid phase ($\kappa \neq 0$), $\mathcal{R}(t)$ and $\mathcal{H}(t)$ do not commute with the matrix associated with the particle number, $\mathcal{N} \equiv \begin{pmatrix} 1 & 0 \\ 0 & -1 \end{pmatrix}$. Thus, we may construct the generalized density $R(t)$ defined in the frame of reference rotating in the gauge space.

$$R(t) = \exp(i\mu\mathcal{N}t)\mathcal{R}(t)\exp(-i\mu\mathcal{N}t) = \begin{pmatrix} \rho(t) & \kappa(t)e^{2i\mu t} \\ -\kappa^*(t)e^{-2i\mu t} & 1 - \rho^*(t) \end{pmatrix}. \tag{16}$$

For this $R(t)$, TDHFB equation (11) becomes

$$i\frac{\partial}{\partial t}R(t) = [H(t) - \mu\mathcal{N}, R(t)], \tag{17}$$

where

$$H(t) = \exp(i\mu\mathcal{N}t)\mathcal{H}(t)\exp(-i\mu\mathcal{N}t) = \begin{pmatrix} h(t) & \Delta(t)e^{2i\mu t} \\ -\Delta^*(t)e^{-2i\mu t} & -h^*(t) \end{pmatrix} \tag{18}$$

Namely, the transformation does not change ρ and $h(t)$, but modifies $\kappa(t)$ and $\Delta(t)$ by multiplying the time-dependent phase $e^{2i\mu t}$. As far as $\Delta(t)$ linearly depends on $\kappa(t)$ and $h(t)$ is a functional of a product of $\kappa(t)$ and $\kappa^*(t)$, $H[R(t)]$ has the functional form identical to $\mathcal{H}[\mathcal{R}(t)]$. The stationary state can be defined by the time-independent density in the rotating frame, $\partial R/\partial t = 0$, leading to

$$[H - \mu\mathcal{N}, R] = 0, \tag{19}$$

where μ correspond to the chemical potential that should be determined by the condition on the particle number. With a proper choice of the $U(M)$ degrees of freedom, the quasiparticle orbitals Φ_ν become time-independent eigenstates as $(H - \mu\mathcal{N})\Phi_\nu = E_\nu\Phi_\nu$. The collective representations, $2M \times M$ matrices Φ and $\bar{\Phi}$, are constructed in the same manner as Eq. (14).

The HFB state is a *quasi-stationary* state in which the pair density and potential are *rotating* in the complex plain with a constant angular velocity 2μ . This is a collective motion corresponding to the Nambu-Goldstone mode associated with the spontaneous breaking of the gauge symmetry, called "pair rotation".

Journal of Physics: Conference Series **387** (2012) 012015 doi:10.1088/1742-6596/387/1/012015

4. Finite amplitude method

Although there have been significant efforts to develop a numerical solver of the TDHFB equation in last decade [22, 23], the computation of the full three-dimensional (3D) dynamics is still a difficult and challenging task. A possible simplification is given by the small-amplitude approximation. It is well-known that this will lead to the quasiparticle random-phase approximation (QRPA). Recently, we have proposed a feasible approach for the linear response calculation [24] and it has been extended for the QRPA [25]. The method is called "finite amplitude method" (FAM).

The QRPA equations can be written in a form

$$\begin{aligned}(E_\nu + E_{\nu'} - \omega)X_{\nu\nu'}(\omega) + (H_1)^{20}_{\nu\nu'}(\omega) &= -(\mathcal{V}_1)^{20}_{\nu\nu'}(\omega),\\ (E_\nu + E_{\nu'} + \omega)Y_{\nu\nu'}(\omega) + (H_1)^{02}_{\nu\nu'}(\omega) &= -(\mathcal{V}_1)^{02}_{\nu\nu'}(\omega),\end{aligned} \tag{20}$$

where

$$(H_1)^{20}_{\nu\nu'}(\omega) = \left[\Phi^\dagger H_1(\omega)\bar{\Phi}\right]_{\nu\nu'}, \quad (\mathcal{V}_1)^{20}_{\nu\nu'}(\omega) = \left[\Phi^\dagger \mathcal{V}_1(\omega)\bar{\Phi}\right]_{\nu\nu'}, \tag{21}$$

$$(H_1)^{02}_{\nu\nu'}(\omega) = -\left[\bar{\Phi}^\dagger H_1(\omega)\Phi\right]_{\nu\nu'}, \quad (\mathcal{V}_1)^{02}_{\nu\nu'}(\omega) = -\left[\bar{\Phi}^\dagger \mathcal{V}_1(\omega)\Phi\right]_{\nu\nu'}. \tag{22}$$

Here, $\mathcal{V}_1(\omega)$ is an external field and $H_1(\omega)$ is an induced residual field. The most tedious part is the calculation of $H_1(\omega)$ expanded in terms of $X(\omega)$ and $Y(\omega)$. The FAM calculates the induced residual fields as follows [25]:

$$H_1^{20}(\omega) = \Phi^\dagger \frac{H[R_\eta] - H[R_0]}{\eta}\bar{\Phi}, \qquad H_1^{02}(\omega) = -\bar{\Phi}^\dagger \frac{H[R_\eta] - H[R_0]}{\eta}\Phi. \tag{23}$$

where η is a small real parameter. Here, $R_\eta(\omega) = \bar{\Psi}'_\eta(\omega)\bar{\Psi}^\dagger_\eta(\omega)$, with

$$\bar{\Psi}'_\eta(\omega) = \bar{\Phi} + \eta\Phi X(\omega), \qquad \bar{\Psi}^\dagger_\eta(\omega) = \left(\bar{\Phi} + \eta\Phi Y^*(\omega)\right)^\dagger. \tag{24}$$

Equivalently, the FAM formula can be written in terms of the qp orbitals as

$$H_1^{20}(\omega) = \Phi^\dagger \frac{H[\bar{\Psi}'_\eta, \bar{\Psi}^\dagger_\eta] - H[\bar{\Phi}, \bar{\Phi}^\dagger]}{\eta}\bar{\Phi}, \qquad H_1^{02}(\omega) = -\bar{\Phi}^\dagger \frac{H[\bar{\Psi}'_\eta, \bar{\Psi}^\dagger_\eta] - H[\bar{\Phi}, \bar{\Phi}^\dagger]}{\eta}\Phi. \tag{25}$$

According to these FAM formulae, the residual fields can be calculated by using the finite difference with the parameter η. In this way, the calculation requires only the HFB Hamiltonian $H[R]$.

Computer programs of the FAM have been developed for spherical nuclei [25] based on the HFBRAD [26], and for axially deformed nuclei [27] based on the HFBTHO [28]. The FAM in the 3D grid-space representation was also achieved [29, 30] for nuclei in the normal phase. Here, we show the isoscalar monopole strength function in ^{124}Sn in Fig. 1. The ground state is obtained by the HFBRAD using the SkM* functional and the volume-type pairing. The zero-range nature of the pairing functional requires us to truncate the model space. The truncation was done according to the cutoff for the qp energy, and results with different cutoff energies are shown in Fig. 1 as well. The final results are almost independent from the choice of the cutoff.

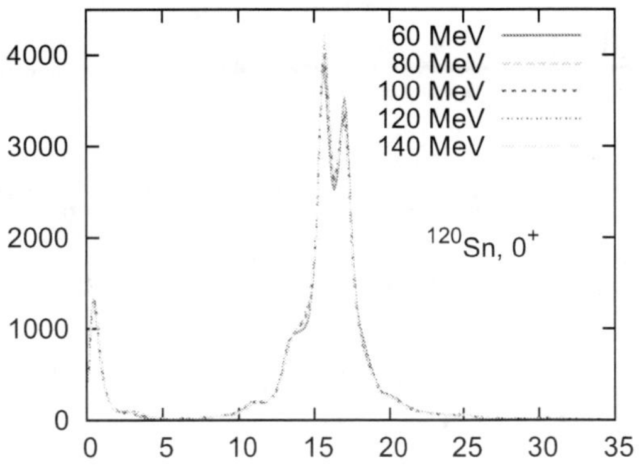

Figure 1. Isoscalar monopole strength function in ^{120}Sn. See text for details.

Journal of Physics: Conference Series **387** (2012) 012015 doi:10.1088/1742-6596/387/1/012015

5. Canonical-basis TDHFB

In this section, we present another approximate treatment of the TDHFB. Here, we do not take the small amplitude limit, instead, assume the diagonal property of the pair potential. In fact, in the stationary limit, this corresponds to the the well-known BCS approximation for the ground state [15]. Therefore, it can be regarded as the BCS-like approximation in the time-dependent treatment.

At every instant of time, we may identify the canonical basis in which the density is diagonal, $\rho_{ij}(t) = \rho_i(t)\delta_{ij}$. The canonical states $(i, \bar{i})$ are always paired with $\rho_{ii}(t) = \rho_{\bar{i}\bar{i}}(t) = \rho_i(t)$ and $\kappa_{i\bar{j}}(t) = \kappa_i(t)\delta_{ij}$. Here, we introduce an assumption that the pair potential is also diagonal in the canonical basis, $\Delta_{i\bar{j}} = -\Delta_i(t)\delta_{ij}$ In the stationary limit, this approximation is identical to the usual BCS approximation [15]. Then, we end up the following set of equations [31].

$$i\frac{\partial}{\partial t}|i(t)\rangle = \{h(t) - \eta_i(t)\}\,|i(t)\rangle, \qquad i\frac{\partial}{\partial t}|\bar{i}(t)\rangle = \{h(t) - \eta_{\bar{i}}(t)\}\,|\bar{i}(t)\rangle, \tag{26}$$

$$i\frac{d}{dt}\rho_i(t) = \kappa_i(t)\Delta_i^*(t) - \kappa_i^*(t)\Delta_i(t), \tag{27}$$

$$i\frac{d}{dt}\kappa_i(t) = \{\eta_i(t) + \eta_{\bar{i}}(t)\}\,\kappa_i(t) + \Delta_i(t)\,\{2\rho_i(t) - 1\}. \tag{28}$$

Here, $\eta_i(t)$ and $\eta_{\bar{i}}(t)$ are arbitrary real functions to control the gauge degrees of freedom of the canonical states. Equations (26), (27), and (28) are invariant with respect to the gauge transformation with arbitrary real functions, $\theta_i(t)$ and $\theta_{\bar{i}}(t)$.

$$|\phi_i\rangle \to e^{i\theta_i(t)}|\phi_i\rangle \quad \text{and} \quad |\phi_{\bar{i}}\rangle \to e^{i\theta_{\bar{i}}(t)}|\phi_{\bar{i}}\rangle \tag{29}$$

$$\kappa_k \to e^{-i(\theta_i(t)+\theta_{\bar{i}}(t))}\kappa_i \quad \text{and} \quad \Delta_k \to e^{-i(\theta_i(t)+\theta_{\bar{i}}(t))}\Delta_i \tag{30}$$

simultaneously with

$$\eta_i(t) \to \eta_i(t) + \frac{d\theta_i}{dt} \quad \text{and} \quad \eta_{\bar{i}}(t) \to \eta_{\bar{i}}(t) + \frac{d\theta_{\bar{i}}}{dt}. \tag{31}$$

Thus, $\eta_i(t)$ and $\eta_{\bar{i}}(t)$, control time evolution of the phases of $|\phi_i(t)\rangle$, $|\phi_{\bar{i}}(t)\rangle$, $\kappa_i(t)$, and $\Delta_i(t)$.

The canonical-basis TDHFB equations (26), (27), and (28), with a proper gauge choice guarantee the following properties [31]:

- (i) Conservation law
  - (a) Conservation of the orthonormal property of canonical states
  - (b) Conservation of the average particle number
  - (c) Conservation of the average total energy
- (ii) The stationary solution corresponds to the HF+BCS state.
- (iii) In the small-amplitude limit, the Nambu-Goldstone modes correspond to zero-energy normal-mode solutions.

In this study, although the method is applicable to the large amplitude dynamics, it is utilized to study the photoreaction of Xe isotopes in the linear regime. Using the instantaneous perturbative external $E1$ field, we calculate the time evolution in the 3D grid-space representation. Then, the Fourier transform of the time-dependent $E1$ moment leads to the $E1$ response function. The smoothing parameter of $\Gamma = 1$ MeV is used. See Ref. [31] for numerical details. The calculated $E1$ strength distributions in $^{132-140}$Xe are shown in Fig. 2.

Journal of Physics: Conference Series **387** (2012) 012015 doi:10.1088/1742-6596/387/1/012015

The ground states of these nuclei are calculated to be spherical, except for ^{140}Xe which has a very small deformation of $\beta = 0.02$. The peak energy of the giant dipole resonance (GDR) is roughly constant and $E_{\mathrm{GDR}} \approx 15$ MeV for these isotopes. ^{136}Xe corresponds to the neutron magic number $N = 82$. The property of the GDR does not significantly depend on the neutron magicity. In the low-energy region below $E = 10$ MeV, we may notice an onset of a small dipole peak beyond $N = 82$, which appear in ^{138}Xe and increases in ^{140}Xe. This seems to be mainly due to a drastic decrease in the neutron separation energy beyond $N = 82$. A systematic study on the low-energy $E1$ strength in this mass region is currently under progress.

It should be emphasized that the computational cost of the present real-time approach is significantly smaller than the normal QRPA calculation based on the diagonalization method. For instance, the computational cost of the QRPA calculation in Ref. [32] is larger than the present one by several orders of magnitude, even though the axial symmetry restriction was utilized. This is because the canonical-basis method utilize the selected canonical states whose number is much smaller than the number of qp states. In addition, the real-time method is very efficient for the present purpose, because the single time evolution produces the nuclear response for the entire energy region.

6. Summary

We presented basic concepts of the density functional theory in nuclear physics, based on the historical developments in nuclear many-body theories. The energy density functional is established as the density-dependent effective interactions, which gives a consistent mean-field-type description of the nuclear saturation property. The time-dependent density functional approach is a powerful method to study dynamics of the quantum many-body systems. In description of heavy open-shell nuclei, the Kohn-Sham orbitals should be extended to the Bogoliubov-type quasiparticle orbitals, to include the pair correlations. This is essentially identical to the TDHFB theory. Although the full treatment of the TDHFB is still a challenging task, we presented approximated treatments; the finite amplitude method and the canonical-basis TDHFB method. Both methods serve as an efficient computational approach to dynamical properties of heavy superfluid nuclei.

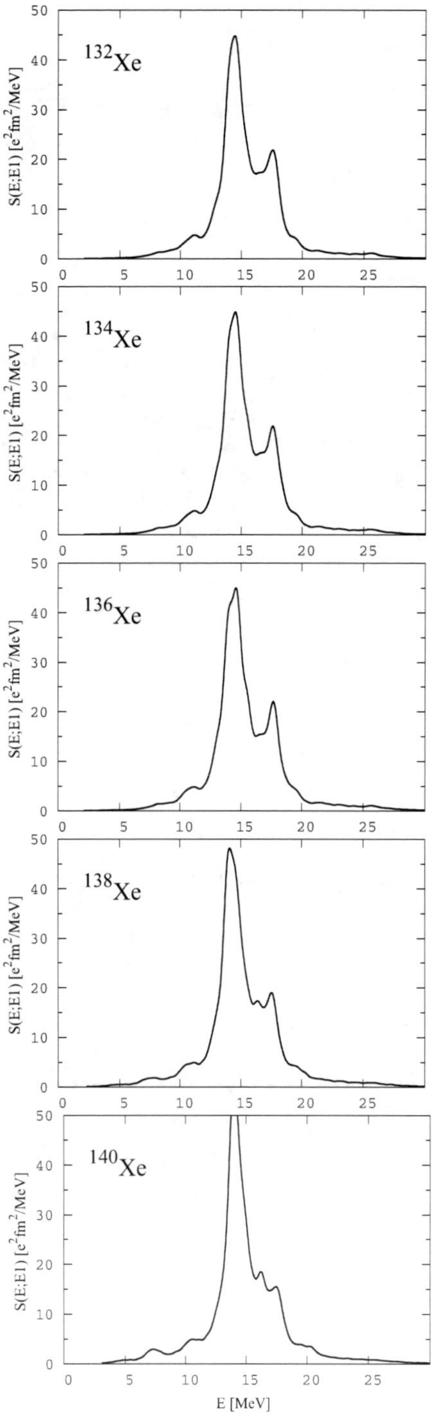

Figure 2. $E1$ strength distributions in Xe isotopes calculated with the SkM* functional.

Acknowledgements
This work was supported by Grant-in-Aid for Scientific Research(B) No. 21340073 and Innovative Areas No. 20105003. The numerical calculations were performed in part on the RIKEN Integrated Cluster of Clusters (RICC), on PACS-CS and T2K in Center for Computational Sciences, University of Tsukuba, and on HITACHI SR16000 in Yukawa Institute, Kyoto University.

References
[1] Bohr A and Mottelson B R 1969 *Nuclear Structure* vol. 1 (New York; W. A. Benjamin)
[2] Pieper S C and Wiringa R B 2001 *Ann. Rev. Nucl. Part. Sci.* **51** 53
[3] Bender M, Heenen P H and Reinhard P-G 2003 *Rev. Mod. Phys.* **75** 121
[4] Lunney D, Pearson J M and Thibault C 2003 *Rev. Mod. Phys.* **75** 1021
[5] Hohenberg P and Kohn W 1964 *Phys. Rev.* **B864** 136
[6] Kohn W and Sham L J 1965 *Phys. Rev.* **A1133** 140
[7] Engel J 2007 *Phys. Rev. C* **75** 014306
[8] Giraud B 2008 *Phys. Rev. C* **78** 014307
[9] Giraud B, Jennings B K and Barrett B R 2008 *Phys. Rev. A* **78** 032507
[10] Weizsäcker C F V 1935 *Z. Phys.* **96** 431
[11] Bethe H A and Bacher R F 1936 *Rev. Mod. Phys.* **8** 82
[12] Mayer M G and Jensen J H D 1955 *Elementary theory of nuclear shell structure* (New York: John Wiley & Sons)
[13] Weisskopf V F 1957 *Nucl. Phys.* **3** 423
[14] Day B D 1967 *Rev. Mod. Phys.* **39** 495
[15] Ring p and Schuck P 1980 *The nuclear many-body problem* (New York; Springer-Verlag)
[16] Bethe H A and Goldstone J 1957 *Proc. Roy. Soc. London* **A238** 551
[17] Day B D 1981 *Phys. Rev. Lett.* **47** 226
[18] Negele J W 1970 *Phys. Rev. C* **1** 1260
[19] Negele J W and Vautherin D 1972 *Phys. Rev. C* **5** 1472
[20] Vautherin D and Brink D M 1972 *Phys. Rev. C* **5** 626
[21] Blaizot J-P and Ripka G 1986 *Quantum Theory of Finite Systems* (Cambridge; MIT Press)
[22] Hashimoto Y and Nodeki K 2007 *Preprint: arXiv:0707.3083*
[23] Stetcu I, Bulgac A, Magierski P and Roche K J 2011 *Phys. Rev. C* **84** 051309
[24] Nakatsukasa T, Inakura T and Yabana K 2007 *Phys. Rev. C* **76** 024318
[25] Avogadro P and Nakatsukasa T 2011 *Phys. Rev. C* **84** 014314
[26] Bennaceur K and Dobaczewski J 2005 *Comp. Phys. Comm.* **168** 96
[27] Stoitsov M, Kortelainen M, Nakatsukasa T, Losa C and Nazarewicz W 2011 *Phys. Rev. C* **84** 041305
[28] Stoitsov M V, Dobaczewski J, Nazarewicz W and Ring P 2005 *Comp. Phys. Comm.* **167** 43
[29] Inakura I, Nakatsukasa T and Yabana K 2009 *Phys. Rev. C* **80** 044301
[30] Inakura I, Nakatsukasa T and Yabana K 2011 *Phys. Rev. C* **84** 021302
[31] Ebata S, Nakatsukasa T, Inakura T, Yoshida K, Hashimoto Y and Yabana K 2010 *Phys. Rev. C* **82** 034306
[32] Terasaki J and Engel J 2010 *Phys. Rev. C* **82** 034326

Journal of Physics: Conference Series **387** (2012) 012016 doi:10.1088/1742-6596/387/1/012016

Ab initio No-core Shell Model Calculations in a SU(3)-based Coupling Scheme

T Dytrych[1], K D Launey[1], J P Draayer[1], D Langr[2]

[1]Department of Physics and Astronomy, Louisiana State University, Baton Rouge, LA 70803, USA

[2] Department of Computer Systems, Czech Technical University in Prague, Prague, Czech Republic

E-mail: tdytrych@phys.lsu.edu

Abstract. We use powerful algorithms of computational group theory to perform *ab initio* configuration-interaction calculations in a SU(3)-based symmetry-adapted many-particle basis. We demonstrate that eigenfunctions for the low-lying states of ^{6}Li, ^{8}Be, ^{12}C, and ^{16}O exhibit a strong dominance of low proton, neutron, and total intrinsic spins that carry the same spatial deformation as the leading symplectic Sp$(3,\mathbb{R})$ irreducible representations. Our findings imply that only a small fraction of the complete model space is needed to model nuclear collective dynamics, deformation, and α-particle clustering even if one uses modern realistic interactions that do not preserve SU(3) symmetry. This in turn points to the importance of using a symmetry-adapted framework, one based on a LS coupling scheme with the associated spatial configurations organized according to deformation.

1. Introduction

Theoretical advances achieved in recent years in the development of realistic nuclear potential models [1, 2, 3, 4] along with progress in high performance computing have placed *ab initio* many-particle approaches [5, 6, 7] at the frontier of nuclear structure explorations. The *ab initio* methods are built on fundamental principles and therefore hold promise to provide predictive capabilities essential for a description of the structure and reactions of unstable and exotic nuclei, many of which are of high interest, e.g. in nucleosynthesis, but remain inaccessible even to experiment.

The no-core shell model (NCSM) [5] is a prominent *ab initio* method that has achieved a good description of low-lying states and associated spectroscopic properties up through p-shell nuclei [8, 9, 10]. The NCSM typically employs the Lanczos algorithm to solve the eigenvalue problem for a realistic Hamiltonian. Matrix elements of the Hamiltonian are calculated in a many-particle basis of m-scheme states, which are constructed as an antisymmetrized product of the harmonic oscillator single-particle wave functions, and carry the z-component of the total angular momentum along with the total parity as good quantum numbers. The main limitation of this approach, and the predictive power thereof, is inherently coupled with the combinatorial growth in the size of the many-particle model space with increasing nucleon numbers and expansion in the number of single-particle levels in the model space.

We developed an innovative *ab initio* model, the symmetry-adapted no-core shell model (SA-NCSM), which utilizes a many-particle basis that exploits the physically relevant SU(3)$\supset$SO(3)

Journal of Physics: Conference Series **387** (2012) 012016 doi:10.1088/1742-6596/387/1/012016

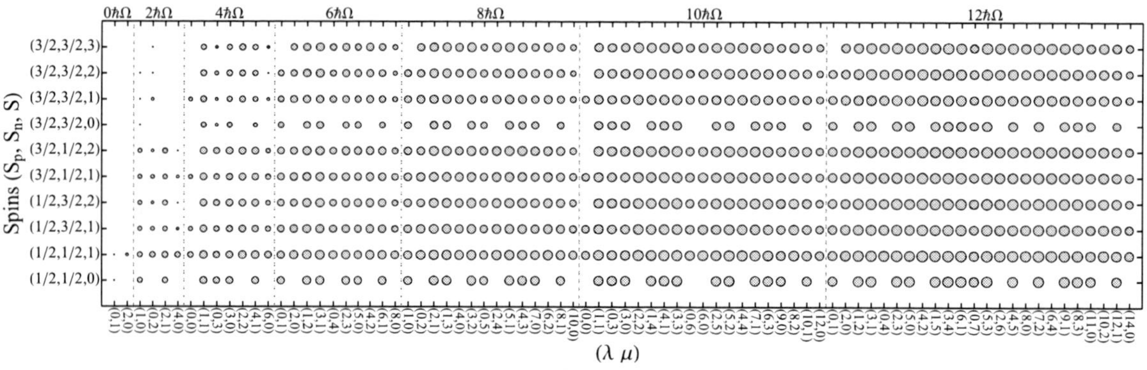

Figure 1. Pauli-allowed deformations $(\lambda\,\mu)$ and their proton, neutron, and total intrinsic spins (S_p, S_n, S) for many-particle SU(3)-scheme configurations of ^{6}Li up through $N_{\max} = 12$. Each circle represents basis states carrying the same $S_pS_nS(\lambda\,\mu)$ quantum numbers, with the radius being proportional to $\log_{10}$ of the number of such states.

group-subgroup chain. The significance of the SU(3) group for a microscopic description of the nuclear collective dynamics can be readily seen from the fact that it is the symmetry group of the Elliott model [11], and a subgroup of the Sp(3, $\mathbb{R}$) symplectic model of nuclear collective motion [12, 13]. The concept of symmetry-adapted many-particle basis represents a powerful tool that allows one to winnow a model space to correlations indispensable for modeling important modes of nuclear collective dynamics, specifically nuclear deformation and cluster substructures, thereby overcoming the scale explosion bottleneck of *ab initio* nuclear structure computations. Hence, SA-NCSM framework holds promise to expand dramatically the reach of current *ab initio* approaches toward describing heavier mass nuclei with unprecedented accuracy.

2. *Ab initio* Calculations in a SU(3) Symmetry-Adapted Basis

The basis states of the SA-NCSM are constructed in the proton-neutron formalism and are labeled by the physical SU(3)⊃SO(3) subgroup chain quantum numbers $(\lambda\,\mu)\kappa L$, and by proton, neutron, and total intrinsic spins S_p, S_n, and S. The orbital angular momentum L is coupled with S to the total angular momentum J and its projection M_J. Each basis state is thus labeled in the SU(3)-scheme as

$$|\vec{\alpha}\,(\lambda\,\mu)\kappa L; S_pS_nS; JM_J\rangle, \tag{1}$$

where the deformation-related $(\lambda\,\mu)$ set of quantum numbers labels SU(3) irreducible representations (irreps) and bring forward important information about nuclear shapes and deformation. For example, (00), $(\lambda\,0)$ and $(0\,\mu)$ describe spherical, prolate and oblate shapes, respectively. The label κ distinguishes multiple occurrences of the same L value in the parent irrep $(\lambda\,\mu)$. The symbol $\vec{\alpha}$ schematically denotes the additional quantum numbers needed to unambiguously distinguish between irreps carrying the same $S_pS_nS(\lambda\,\mu)$ quantum numbers. These irreps compose a well-defined subspace with a unique feature that allows for the complete separation of intrinsic and center-of-mass degrees of freedom. The size of these subspaces is typically several orders of magnitude smaller than the full $N\hbar\Omega$ space, where the m-scheme basis also allows for an exact factorization of center-of-mass degrees of freedom. The decomposition of the SA-NCSM model space of ^{6}Li up through $N_{\max} = 12$ model space is depicted in Fig. 1.

The SA-NCSM implements a set of powerful algorithms [14, 15], which facilitate calculations

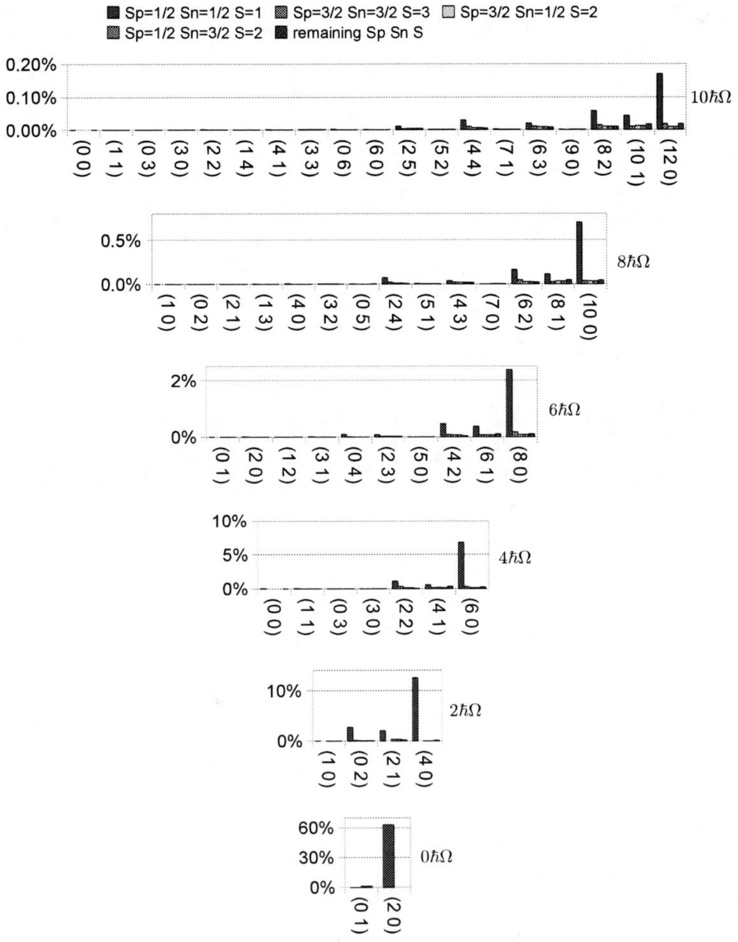

Figure 2. Probability distribution of proton, neutron, and total intrinsic spin components (S_p, S_n, S) across their Pauli-allowed spatial deformations ($\lambda\,\mu$) [horizontal axis] in the ground 1^+ state of ^{6}Li obtained for $N_{\max} = 10$ and $\hbar\Omega = 20$ MeV.

of matrix elements of arbitrary (currently up to two-body, but expandable to higher-rank) operators in the SU(3)-scheme basis. This allows for both the evaluation of the Hamiltonian matrix elements, and the use of the resulting eigenvectors to evaluate other experimental observables. The underlying principle behind the SA-NCSM computational kernel is a SU(3)-type Wigner-Eckhart theorem, which allows the problem to be factorized into SU(3) reduced matrix elements (*rmes*) and SU(3) coupling/recoupling coefficients. While the latter can be computed using the publicly available library [16], the former is calculated from a set of single-shell *rmes* by the repetitive application of a SU(3)-type reduction formula for *rmes* of operators acting on the two independent, proton and neutron subsystems.

3. Results

We first calculate binding energies and low-lying eigenstates of ^{6}Li, ^{8}Be, ^{12}C, and ^{16}O nuclei using the JISP16 bare nucleon-nucleon (NN) interaction, for a wide range of HO energies $\hbar\Omega$ and the maximal full $N_{\max}$ model spaces attainable by the current implementation of the SA-NCSM approach. The resulting wave functions are then employed to obtain physical observables such as mass rms radii, electric quadrupole moments, magnetic dipole moments, reduced electromagnetic

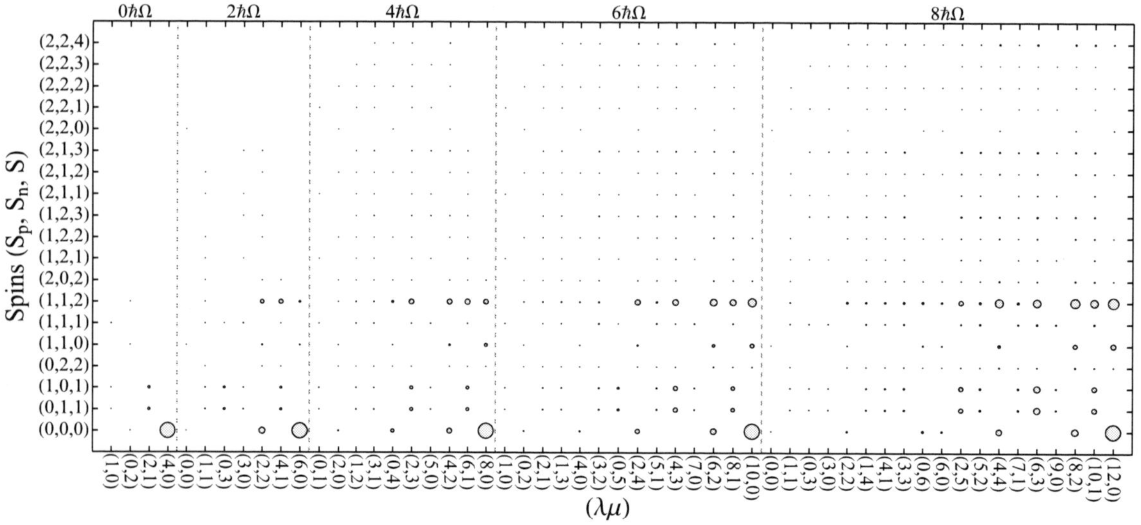

Figure 3. Probability distribution of $(S_p,\, S_n,\, S)$ spin components across their Pauli-allowed spatial deformations $(\lambda\,\mu)$ in the ground 0^+ state of ^{8}Be obtained for $N_{\max} = 8$ and $\hbar\Omega = 22.5$ MeV. The area of each circle is proportional to the total probability of $k\hbar\Omega$ states carrying the same $S_pS_nS(\lambda\,\mu)$ quantum numbers normalized with respect to the total probability of $k\hbar\Omega$ $S_p = 0\, S_n = 0\, S = 0$ $(4 + k\, 0)$ stretched-states.

B(E2) and B(M1) transition strenghts. Direct comparison of the SA-NCSM outcomes with the results of analogous m-scheme based NCSM calculations reveals agreement typically to within the numerical precision of four significant figures, and thereby unambiguously confirming the validity of *ab initio* SA-NCSM approach, and the correctness of our implementation of it.

The expansion of nuclear wave functions in a physically relevant SU(3)-scheme basis is illuminating salient features that emerge from the complex dynamics of strongly interacting many-particle systems. To investigate the nature of the most important many-particle correlations, we enumerate the probability distribution of intrinsic spin $(S_p\, S_n\, S)$ and deformation-related $(\lambda\,\mu)$ quantum numbers for the low-lying $T = 0$ states of ^{6}Li, members of the ground-state rotational band in ^{8}Be and ^{12}C, and the ground state of ^{16}O. Figure 2 shows the probability distribution of intrinsic spins across their Pauli-allowed deformations in the ground state of ^{6}Li. This figure illustrates a facet common to low-energy *ab initio* solutions we study: a hitherto unrecognized simple mixing pattern of intrinsic spins and SU(3) spatial quantum numbers.

We observe that over 99% of the SA-NCSM eigenstates is typically accounted for by a small fraction of intrinsic spin combinations. For instance, the lowest-lying $T = 0$ eigenstates in ^{6}Li are almost entirely realized in terms of configurations characterized by the following intrinsic spin $(S_p\, S_n\, S)$ triplets: $(\frac{3}{2}\,\frac{3}{2}\,3)$, $(\frac{1}{2}\,\frac{3}{2}\,2)$, $(\frac{3}{2}\,\frac{1}{2}\,2)$, $(\frac{1}{2}\,\frac{1}{2}\,0)$ and $(\frac{1}{2}\,\frac{1}{2}\,1)$ with the last carrying over 90% of the eigenstate. Similarly, the ground state band in ^{8}Be and ^{12}C along with the ground state of ^{16}O are dominated by many-particle configurations carrying the total intrinsic spin of protons and neutrons equal to zero and one with the highest probability due to the dominant $S_p = S_n = S = 0$ configurations.

The mixing of $(\lambda\,\mu)$ spatial quantum numbers in the lowest-lying $T = 0$ eigenstates in ^{6}Li, the ground state band in ^{8}Be and ^{12}C along with the ground state of ^{16}O, induced by the SU(3) symmetry breaking terms of realistic interactions, exhibits a remarkably regular pattern. One of its key features is the preponderance of a single $0\hbar\Omega$ SU(3) irrep, the so-called leading irrep, that is, the one characterized by the largest value of the second order SU(3) Casimir invariant, $\hat{C}_2$, and hence corresponding to a large intrinsic quadrupole deformation. For instance, the low-lying $T = 0$ states of ^{6}Li project at 40%-70% level onto the prolate-like $0\hbar\Omega$ SU(3) irrep $(2\,0)$, as illustrated in Fig. 2 for the ground state. Qualitatively similar dominance of the leading $0\hbar\Omega$ SU(3) irreps $(4\,0)$, $(0\,4)$, and $(0\,0)$, associated with prolate, oblate, and spherical shapes is observed for the investigated eigenstates of ^{8}Be, ^{12}C, and ^{16}O, respectively. Such clear prominence of the *p*-shell states with the highest spatial symmetry points to the fact that the effective quadrupole-quadrupole interaction of Elliott SU(3) model of nuclear rotations is realized naturally within the framework of the modern realistic NN interactions.

Our analysis reveals that the dominant SU(3)-scheme states at each $k\hbar\Omega$ subspace are typically those with $(\lambda\,\mu)$ quantum numbers such that

$$\lambda + 2\mu = \lambda_0 + 2\mu_0 + k \qquad k = 2, 4, \ldots, \tag{2}$$

where λ_0 and μ_0 denote labels of a leading SU(3) irrep at the $k = 0$, $0\hbar\Omega$ level. We conjecture that such an intriguing mixing pattern of SU(3) quantum numbers reflects the presence of an underlying symplectic $\mathrm{Sp}(3,\mathbb{R})$ symmetry of microscopic nuclear collective motion that governs the low-energy structure of both even-even and odd-odd *p*-shell nuclei. This can be seen from the fact that the core excitations with $(\lambda\,\mu)$ deformation shapes satisfying condition (2) can be determined from the leading SU(3) irrep $(\lambda_0\,\mu_0)$ through a successive application of a specific subset of the $\mathrm{Sp}(3,\mathbb{R})$ symplectic $2\hbar\Omega$ raising operators. This subset is composed of $\hat{A}_{zz}$, $\hat{A}_{zx}$, and $\hat{A}_{xx}$ operators that distribute two oscillator quanta in z and x directions, but none in y direction, thereby inducing SU(3)-scheme configurations with ever-increasing values of the Casimir invariant $\hat{C}_2$. These three operators are the generators of the $\mathrm{Sp}(2,\mathbb{R}) \subset \mathrm{Sp}(3,\mathbb{R})$ subgroup [17], and give rise to a hierarchy of deformation shapes that are energetically favored by an attractive quadrupole-quadrupole interactions [13]. Furthermore, there is an apparent hierarchy among states that fulfill condition (2). In particular, the $k\hbar\Omega$ configurations with $(\lambda_0 + k,\ \mu_0)$, the so-called stretched-states, carry a very noticeably higher probability than the others , as can be readily seen in Fig. 2 and Fig. 3. For instance, the $(2 + k\,0)$ stretched-states contribute at 85% level to the ground state of ^{6}Li. The sequence of the stretched-states, that is, the states with the highest possible deformations, can be formed from many-nucleon correlations of a leading SU(3) irrep by application of the $\hat{A}_{zz}$ operator, which is the generator of $\mathrm{Sp}(1,\mathbb{R}) \subset \mathrm{Sp}(2,\mathbb{R}) \subset \mathrm{Sp}(3,\mathbb{R})$ subgroup. The important role of the stretched-configurations for the description of the rotational bands in $N = Z$ even-even nuclei of was recognized heretofore using a simplistic microscopic Hamiltonian [18]. Here, for the first time, we establish its presence in the framework of the *ab initio* nuclear structure study.

The revealed pattern of intrinsic spin and deformation mixing supports a symmetry guided truncation of the N_{max} model space. Clearly, one can take advantage of the physical relevance of the SU(3)-scheme basis to winnow the full space down to the most relevant configurations that support the strongest many-nucleon correlations of the system using the underlying $\mathrm{Sp}(1,\mathbb{R}) \subset \mathrm{Sp}(2,\mathbb{R}) \subset \mathrm{Sp}(3,\mathbb{R})$ symmetry considerations. As noted previously, this truncation, while significantly reducing the size of the model space, also preserves the ability to factor out exactly the spurious center-of-mass degrees of freedom.

4. Conclusion

We have developed a novel approach that capitalizes on advances being made in *ab initio* methods while exploiting exact and partial symmetries of nuclear many-body system. Using this approach we have demonstrated that the low-lying eigenstates of ^{6}Li, ^{8}Be, ^{12}C, and ^{16}O, which were obtained using the JISP16 NN interaction, exhibit a strong dominance of few intrinsic spin components and carry an intriguingly simple pattern of dominant deformations. This demonstrates an important role of the symplectic $\mathrm{Sp}(3,\mathbb{R})$ symmetry in light nuclei while reaffirming the value of the simpler SU(3) model upon which it is based. The SA-NCSM thus offers a systematic framework for down-selecting many-body correlations to physically relevant and manageable subspaces based on spin and deformation selection, while preserving the ability to factorize the center-of-mass degrees of freedom exactly.

Acknowledgments
Supported by the U.S. NSF (PHY-0500291 & OCI-0904874), the U.S. DOE (DE-SC0005248), and the SURA. This research used computing resources of the Louisiana Optical Network Initiative, LSU's Center for Computation & Technology, and the National Energy Research Scientific Computing Center, which is supported by the Office of Science of the U.S. Department of Energy under Contract No. DE-AC02-05CH11231.

References
[1] Shirokov A, Vary J, Mazur A and Weber T 2007 *Phys. Lett. B* **644** 33
[2] Wiringa R B, Stoks V G J and Schiavilla R 1995 *Phys. Rev. C* **51** 38
[3] Entem D R and Machleidt R 2003 *Phys. Rev. C* **68** 041001
[4] Epelbaum E 2006 *Prog.Part.Nucl.Phys* **57** 654
[5] Navrátil P, Vary J P and Barrett B R 2000 *Phys. Lett.* **84** 5728
[6] Pieper S C, Varga K and Wiringa R B 2002 *Phys. Rev. C* **66** 044310
[7] Wloch M, Dean D J, Gour J R, Hjorth-Jensen M, Kowalski K, Papenbrock T and Piecuch P 2005 *Phys. Rev. Lett.* **94** 212501
[8] Navrátil P, Quaglioni S, Stetcu I and Barrett B R 2009 *J. Phys. G: Nucl. Part.* **36** 083101
[9] Roth R and Navrátil P 2007 *Phys. Rev. Lett.* **99** 092501
[10] Navratil P, Gueorguiev V, Vary J P, Ormand W and Nogga A 2007 *Phys. Rev. Lett.* **99** 042501 LANL ArXiV: Nucl-Th 0701038
[11] Elliott J P 1958 *Proc. Roy. Sot. A* **245** 128
[12] Rosensteel G and Rowe D 1977 *Phys. Rev. Lett.* **38** 10
[13] Rowe D 1985 *Reports on Progr. in Phys.* **48** 1419
[14] Draayer J P, Leschber Y, Park S C and Lopez R 1989 *Comput. Phys. Commun.* **56** 279
[15] Bahri C and Draayer J 1994 *Comput. Phys. Commun.* **83** 59
[16] Akiyama Y and Draayer J 1973 *Comput. Phys. Commun.* **5** 405
[17] Peteron D R and Hecht K T 1980 *Nucl. Phys. A* **344** 361
[18] Arickx F 1976 *Nucl. Phys. A* **268** 347

Journal of Physics: Conference Series **387** (2012) 012017 doi:10.1088/1742-6596/387/1/012017

Symmetry-Adapted Ab Initio Open Core Shell Model Theory

J P Draayer[1], T Dytrych[1], K D Launey[1], D Langr[2], A C Dreyfuss[3], and C. Bahri[4]

[1]Department of Physics and Astronomy, Louisiana State University, Baton Rouge, LA 70803, USA

[2] Department of Computer Systems, Czech Technical University in Prague, Prague, Czech Republic

[3] Keene State College, Keene, New Hampshire 03435, USA

[4] Department of Physics, University of Notre Dame, Notre Dame, Indiana 46556-5670, USA

E-mail: draayer@lsu.edu

Abstract. By using only a fraction of the model space, we gain further insight – within a symmetry-guided no-core shell model framework – into the many-body nuclear dynamics that gives rise to important single-particle configurations together with correlated highly-deformed and alpha-cluster structures. We show results of the novel *ab initio* symmetry-adapted no-core shell model for large-scale nuclear structure computations. In addition, we use the symmetry patterns unveiled in these results to explore ultra-large model spaces.

1. Introduction

The major progress achieved during the recent years in high performance computing and in the internucleon interactions derived from first principles [1, 2] have placed *ab initio* many-body approaches at the frontier of nuclear structure explorations [3, 4, 5, 6, 7]. The ultimate goal of *ab initio* nuclear structure modeling is to establish a link between the underpinning principles of Quantum Chromodynamics (quark/gluon considerations) and properties of atomic nuclei, and in doing so, to use the predictive capabilities toward providing reliable descriptions of nuclei far off the valley of stability. This, in turn, yields highly accurate nuclear wave functions, which are key ingredients to calculating observables for nuclei and reactions in extreme environments, from the interiors of stars to the core of nuclear reactors.

We present first results of an innovative *ab initio* model, the symmetry-adapted no-core shell model (SA-NCSM) [8, 9]. It utilizes an SU(3)-symmetric basis, which is related – via a unitary transformation – to the three-dimensional harmonic oscillator (HO) many-body basis (m-scheme) used in the no-core shell model (NCSM) [3, 7]. In particular, we show that *ab initio* SA-NCSM calculations for p-shell nuclei in full space (equivalent to the NCSM results) reveal a clear $\mathrm{Sp}(3,\mathbb{R})$ symplectic structure with no *a priori* restrictions to the symplectic symmetry. This, in turn, allows the growth of the model space within the SA-NCSM framework to be managed by winnowing to only physically relevant states as determined through symmetry considerations, thereby overcoming a prime bottleneck of the NCSM – the combinatorial growth of many-body basis with number of nucleons and model space truncation, $N_{\max}$. As a complementary study, we examine the significance of employing ultra-large model spaces, much beyond the current

$N_{\max}$ limit. Specifically, we employ the no-core symplectic model (NCSpM) that utilizes the $\mathrm{Sp}(3,\mathbb{R})$ symplectic basis, which can also be obtained via a unitary transformation from the NCSM m-scheme basis. We start with selected $\mathrm{Sp}(3,\mathbb{R})$ symplectic structures, those that are found most important in the *ab initio* results, and in addition, we use a simple internucleon interaction with only one adjustable parameter. This enables one to accommodate, for the first time, shell-model basis spaces that are believed to be essential to achieve convergence of, e.g., alpha-cluster states. The results for ^{12}C reveal that a large-$N_{\max}$ model space is indeed the key to describe – in a NCSM framework – the 'Hoyle' state, the second 0^+ state in ^{12}C at 7.65 MeV, that is of particular astrophysical interest.

2. *Ab initio* SA-NCSM Results: Symmetry Patterns

The *ab initio* SA-NCSM adopts the first-principle concept and utilizes a many-particle basis that is reduced with respect to the physically relevant $\mathrm{SU}(3)\supset\mathrm{SO}(3)$ subgroup chain. The significance of the SU(3) group for a microscopic description of the nuclear collective dynamics can be seen from the fact that it is the symmetry group of the Elliott model [10, 11, 12], and a subgroup of the $\mathrm{Sp}(3,\mathbb{R})$ symplectic model [13]. By recognizing that the choice of coordinates, especially when deformed nuclear shapes dominate, is crucial, the SA-NCSM affords a solution in terms of coordinates that reflect symmetries inherent to the nuclear system. This allows the full model space to be down-selected to the physically relevant space.

The $C^{(1\,1)}_{\mathcal{L}M}$ translationally invariant (intrinsic) eight generators of the SU(3) group can be written in terms of particle position and momentum coordinates and are related to the angular momentum operator L_{1q} ($= C^{(1\,1)}_{1q}$, $q = 0, \pm1$) and the in-shell quadrupole moment tensor $\mathcal{Q}_{2q}$ ($= \sqrt{3}C^{(1\,1)}_{2q}$, $q = 0, \pm1, \pm2$). The deformation-related $(\lambda\,\mu)$ set of quantum numbers labels SU(3) irreducible representations, irreps. Consequently, SU(3)-symmetric states (and hence symplectic basis states that are built on these) bring forward important information about nuclear shapes and deformation in terms of $(\lambda\,\mu)$, for example, $(0\,0)$, $(\lambda\,0)$ and $(0\,\mu)$ describe spherical, prolate and oblate shapes, respectively.

The conventional NCSM basis spaces are constructed using HO single-particle states and are characterized by the $\hbar\Omega$ oscillator strength as well as by the cutoff in total oscillator quanta, $N_{\max}$, above the lowest energy configuration for a given nucleus. The basis states of the SA-NCSM for a given $N_{\max}$ are constructed in the proton-neutron formalism using also HO single-particle states and are labeled by the $\mathrm{SU}(3)\supset\mathrm{SO}(3)$ subgroup chain quantum numbers $(\lambda\,\mu)\kappa L$, together with proton, neutron, and total intrinsic spins S_π, S_ν, and S. The orbital angular momentum L is coupled with S to the total orbital momentum J and its projection M_J. Each basis state in this scheme is labeled schematically as $|\vec{\gamma}\,(\lambda\,\mu)\kappa L; (S_\pi S_\nu)S; JM_J\rangle$. The label κ distinguishes multiple occurrences of the same L value in the parent irrep $(\lambda\,\mu)$, and $\vec{\gamma}$ distinguishes among configurations carrying the same $(\lambda\,\mu)$ and $(S_\pi S_\nu)S$ labels.

The *ab initio* SA-NCSM results for p-shell nuclei, such as ^{6}Li, ^{8}B, ^{12}C and ^{16}O, reveal a dominance of shapes of large deformation (typically large $|\lambda\ -\mu|$) in the $0\hbar\Omega$ subspace. Furthermore, we find that important SU(3) configurations are then organized into structures with $\mathrm{Sp}(3,\mathbb{R})$ symplectic symmetry. As an example, for the ground state of ^{6}Li calculated using the bare JISP16 realistic interaction [14] for $N_{\max} = 10$ and $\hbar\Omega = 20$ MeV, results indicate a clear dominance of the spin configuration, $\{S_\pi = 1/2, S_\nu = 1/2, S = 1\}$, and the most deformed shapes, $(2\,0)$, $(4\,0)$, $(6\,0)$, $(8\,0)$, $(10\,0)$, and $(12\,0)$ in each $\hbar\Omega$ subspace [15]. In the case of ^{6}Li, the $(2\,0)$ symplectic irrep gives rise to $(0\,2)$, $(2\,1)$ and $(4\,0)$ configurations in the $2\hbar\Omega$ subspace, and those configurations indeed realize the major components of the wavefunction in this subspace. This further confirms the significance of the symplectic symmetry to nuclear dynamics. Similar results are observed for the ground and low-lying excited states of ^{8}B, ^{12}C and ^{16}O. The outcome points to the fact that the relevant model space can be systematically down-selected and further expanded to higher $N_{\max}$. For example, only 11% of the total space dimension is needed to yield 99% of the matter rms radius and the $M1$ moment calculated in full space.

In short, the SA-NCSM advances an extensible microscopic framework for studying nuclear structure and reactions that capitalizes on advances being made in *ab initio* methods while exploiting symmetries – exact and partial, known to dominate the dynamics.

3. NCSpM: Highly-Deformed Structures in Ultra-large Model Spaces

The NCSpM, while selecting the most relevant symplectic configurations, is employed to provide shell model calculations beyond current NCSM limits, namely, up through $N_{\max} = 20$ for ^{12}C, the model space we found sufficient for the convergence of results. Clearly, the NCSpM employed within a full model space up through a given $N_{\max}$, will coincide with the NCSM for the same $N_{\max}$ cutoff. However, in the case of the NCSpM, the symplectic irreps divide the space into 'vertical slices' that are comprised of basis states of a definite deformation $(\lambda\,\mu)$. Hence, the model space can be reduced to only a few important configurations that are chosen among all possible $\mathrm{Sp}(3,\mathbb{R})$ irreps within the $N_{\max}$ model space.

The basis states of an $\mathrm{Sp}(3,\mathbb{R})$ irrep (slice) are built by $2\hbar\Omega$ 1-particle-1-hole (1p-1h) monopole or quadrupole excitations (one particle raised by two shells) over a bandhead together with a smaller $2\hbar\Omega$ 2p-2h correction for eliminating the spurious center-of-mass (CM) motion. The symplectic bandhead is the lowest-weight $\mathrm{Sp}(3,\mathbb{R})$ state, which is defined by the usual requirement that the symplectic lowering operators annihilate it – in analogy to a $|J, M_J = -J\rangle$ state for the case of the SU(2) group of angular momentum J. The bandhead is a SU(3)-coupled many-body state with a given nucleon distribution over the HO shells. The corresponding energy of HO quanta, $N_\sigma\hbar\Omega$, together with the bandhead shape serve to label the symplectic irrep [e.g., for ^{12}C, one of the $\mathrm{Sp}(3,\mathbb{R})$ irrep is $N_\sigma = 24.5$ $(0\,4)$ or, equally, $0\hbar\Omega$ $(0\,4)$]. The basis states within an irrep are, in addition, specified by the number n of HO excitation quanta above N_σ and by the resulting $(\lambda\,\mu)$ deformation of the bandhead due to these excitations [e.g., for the ^{12}C $4\hbar\Omega$ $(12\,0)$ irrep, the $n = 2$ basis states ($2\hbar\Omega$ excitations) include $(14\,0)$, $(12\,1)$, and $(10\,2)$ shapes, and so forth for higher n].

In this study we employ a very simple Hamiltonian with an effective interaction derived from the long-range expansion of the two-body central nuclear force,

$$H_{\text{eff}} = H_0 - \frac{\chi}{2}\frac{1}{\gamma}\left(e^{\gamma(Q.Q-\langle Q.Q\rangle_n)} - 1\right), \tag{1}$$

which includes the spherical HO potential (which together with the kinetic energy yields the HO Hamiltonian, H_0) and the $Q.Q$ quadrupole-quadrupole interaction not restricted to a single shell. For the latter term, the average contribution, $\langle Q.Q\rangle_n$, of $Q.Q$ within a subspace of n HO excitations is removed [16], that is, the trace of $Q.Q$ divided by the space dimension for a fixed n. Hence, the large monopole contribution of the $Q.Q$ interaction is removed, which, in turn, helps eliminate the considerable renormalization of the zero-point energy, while retaining the $Q.Q$-driven behavior of the wavefunctions. This Hamiltonian in its zeroth-order approximation (for parameter $\gamma \to 0$) and for a valence shell goes back to the established Elliott model [10, 12]. We take the coupling constant $\chi = \hbar\Omega/(4\sqrt{N_f N_i})$ with N for the initial (i) and final (f) many-body states, where the quantum number $N = N_\sigma + n$ is the total number of HO excitations. The decrease of χ with N, to leading order in λ/N, has been shown by Rowe [17] based on self-consistent arguments. It is important to note that such a model Hamiltonian (1) fixes the strength of the $Q.Q$-term by the value of $\hbar\Omega$, thereby rendering the eigenstates $\hbar\Omega$-independent.

As the interaction and the model space are carefully selected to reflect the most relevant physics, the outcome reveals a quite remarkable agreement with the experiment. The low-lying energy spectrum and eigenstates for ^{12}C were calculated using the NCSpM with H of Eq. (1)

Journal of Physics: Conference Series **387** (2012) 012017 doi:10.1088/1742-6596/387/1/012017

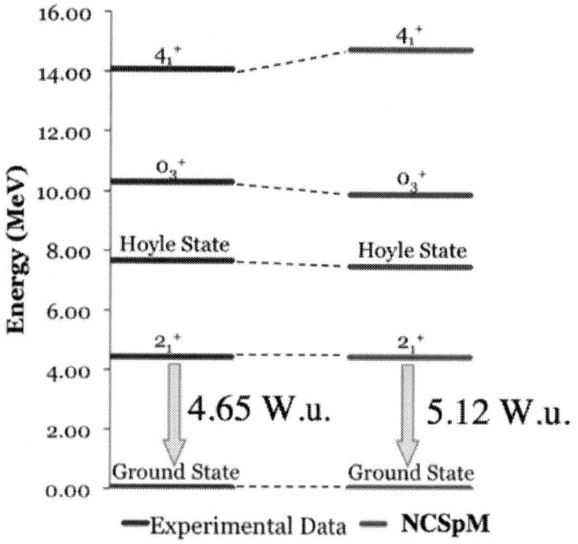

Figure 1. ^{12}C energy spectrum calculated by the NCSpM (right) and compared to experiment (left). The NCSpM calculations are performed using Eq. (1) with $\gamma = -1.71 \times 10^{-4}$ and three $\mathrm{Sp}(3,\mathbb{R})$ irreps, $0\hbar\Omega$ 0p-0h (0 4), $2\hbar\Omega$ 2p-2h (6 2), and $4\hbar\Omega$ 4p-4h (12 0), in the $N_{\mathrm{max}} = 20$ model space.

for $\hbar\Omega = 18$ MeV given by the empirical estimate $\approx 41/A^{1/3} = 17.9$ MeV. The results are shown for $N_{\mathrm{max}} = 20$, which we found sufficient to yield convergence. This N_{max} model space is further reduced by selecting the most relevant symplectic irreps, namely, the spin-zero ($S = 0$) $0\hbar\Omega$ 0p-0h (0 4), $2\hbar\Omega$ 2p-2h (6 2), and $4\hbar\Omega$ 4p-4h (12 0) symplectic bandheads together with all multiples thereof up through $N_{\mathrm{max}} = 20$ of total dimensionality of 4.5×10^3. In comparison to the experimental energy spectrum (Fig. 3), the outcome reveals that the lowest 0^+, 2^+, and 4^+ states of the $0\hbar\Omega$ 0p-0h (0 4) symplectic irrep calculated for $\gamma = -1.71 \times 10^{-4}$ closely reproduce the *g.st.* rotational band, while the calculated lowest 0^+ states of the $4\hbar\Omega$ 4p-4h (12 0) and the $2\hbar\Omega$ 2p-2h (6 2) slices are found to lie close to the Hoyle state and the 10-MeV 0^+ resonance (third 0^+ state), respectively. It is interesting to point out that, for small N_{max}, the 0^+ state of (6 2) appears much lower in energy compared to the 0^+ state of (12 0). However, the latter starts to dramatically drop as N_{max} increases and aligns with the 0^+ state of (6 2) around $N_{\mathrm{max}} = 12$, falling further down for larger N_{max}. Clearly, the mechanism of lowering the highly-deformed (12 0) emerges from the fact that sufficiently large model spaces allow for the collectivity to develop, thereby, making the 0^+ state of (12 0) energetically favorable.

We note that the energy spectrum, calculated in the NCSpM and shown in Figure 3, is rescaled by an overall factor of ~ 2. This factor is determined by fixing the lowest 2^+ by its experimental value. This, however, has no implications on the underlying physics, as an overall factor for H does not change its properties and eigenstates, together with associated observables. For example, the $B(E2;\, 2_1^+ \rightarrow 0_{g.st.}^+)$ is calculated to be 5.12 W.u., also in a close agreement with the 4.65 W.u. experimental measurement. While the model includes an adjustable parameter, γ, this parameter only controls the decrease rate of the $Q.Q$ interaction with increasing n. The entire many-body apparatus is fully microscopic and no adjustments are possible. Hence, as γ varies, there is only a small window of possible γ values that, for large enough N_{max}, closely reproduces the relative positions of the three lowest 0^+ states. As for the ground-state rotational band, the γ parameter has almost no effect on the energies and transition rates.

In short, the present microscopic study with a schematic many-nucleon interaction shows how both collective and cluster-like structures emerge out of a no-core shell-model framework, which can extend to and take into account essential high-lying shell-model configurations.

Journal of Physics: Conference Series **387** (2012) 012017 doi:10.1088/1742-6596/387/1/012017

4. Conclusion

We have presented first results of the SA-NCSM, a fully microscopic theory designed to accept QCD-inspired interactions. We are beginning to observe collective features in the structure of the calculated eigenstates, but not always at the excitation energies found in nature. We have also observed that a simple part of the long-range nuclear force is sufficient to bring such states to the measured energies provided that high-lying shell-model configurations are taken into account. Hence, we are well-positioned to link to QCD considerations, but also to begin to address in a focused way the origin of collective degrees of freedom in nuclei and the underlying physics, that is, the emergence of collectivity from fundamental principles.

Acknowledgments

This work was supported by the U.S. NSF (PHY-0500291 & OCI-0904874), the U.S. DOE (DE-SC0005248), and the SURA; also, partially supported by the U.S. DOE DE-FG02-95ER-40934. ACD acknowledges support by the U.S. NSF (grant 1004822) through the REU Site in Dept. of Physics & Astronomy at LSU. We acknowledge NERSC and LONI for providing HPC resources.

References

[1] Entem D R and Machleidt R 2003 *Phys. Rev. C* **68** 041001
[2] Epelbaum E 2006 *Prog.Part.Nucl.Phys* **57** 654
[3] Navrátil P, Vary J P and Barrett B R 2000 *Phys. Lett.* **84** 5728
[4] Pieper S C, Varga K and Wiringa R B 2002 *Phys. Rev. C* **66** 044310
[5] Wloch M, Dean D J, Gour J R, Hjorth-Jensen M, Kowalski K, Papenbrock T and Piecuch P 2005 *Phys. Rev. Lett.* **94** 212501
[6] Roth R and Navrátil P 2007 *Phys. Rev. Lett.* **99** 092501
[7] Navrátil P, Quaglioni S, Stetcu I and Barrett B R 2009 *J. Phys. G: Nucl. Part.* **36** 083101
[8] Dytrych T, Sviratcheva K D, Bahri C, Draayer J P and Vary J P 2007 *Phys. Rev. Lett.* **98** 162503
[9] Dytrych T, Sviratcheva K D, Draayer J P, Bahri C and Vary J P 2008 *J. Phys. G: Nucl. Part. Phys.* **35** 123101
[10] Elliott J P 1958 *Proc. Roy. Sot. A* **245** 128
[11] Elliott J P 1958 *Proc. Roy. Sot. A* **245** 562
[12] Elliott J P and Harvey M 1962 *Proc. Roy. Sot. A* **272** 557
[13] Rosensteel G and Rowe D 1977 *Phys. Rev. Lett.* **38** 10
[14] Shirokov A, Vary J, Mazur A and Weber T 2007 *Phys. Lett.* **B 644** 33
[15] Dytrych T *et al.* 2012 *submitted*
[16] Rosensteel G and Draayer J P 1985 *Nucl. Phys. A* **436** 445
[17] Rowe D J 1967 *Phys. Rev.* **162** 866

Systematic study of isoscalar and isovector pairing in the $2p1f$ shell

Y Lei[1,2], S Pittel[2], N Sandulescu[3], A Poves[4], B Thakur[2,5] and Y M Zhao[1]

[1] Department of Physics, Shanghai Jiao Tong University, Shanghai, 200240, China
[2] Bartol Research Institute and Department of Physics and Astronomy, University of Delaware, Newark, Delaware 19716, USA
[3] Institute of Physics and Nuclear Engineering, 76900 Bucharest, Romania
[4] Departamento de Fisica Teorica and IFT UAM/CSIC, Universidad Autónoma de Madrid, 28049, Madrid, Spain
[5] LONI Institute and Center for Computational Technology, Louisiana State University, Baton Rouge, LA 70803-4001 USA

E-mail: pittel@bartol.udel.edu

Abstract. A systematic study of isoscalar and isovector pairing in $2p1f$ shell nuclei is reported, based on a model that includes deformation, isoscalar and isovector pairing and spin-orbit effects. Selected results are presented for three $N = Z$ nuclei, ^{44}Ti, ^{46}V and ^{48}Cr, with a focus on the role of single-particle effects in dictating the competing pairing modes.

1. Introduction

In this paper, we review recent work [1] carried out to study the interplay of isoscalar and isovector pairing correlations in atomic nuclei with $N \approx Z$. Our analysis is carried out in the context of the nuclear shell model, which permits us to address the competition between these different modes of pairing in the presence of nuclear deformation and with all symmetries preserved.

The structure of the paper is as follows. In Section 2, we provide a brief historical overview of pairing in nuclei, including not only the usual nn and pp pairing that dominate in heavy systems but also pn pairing as is in principle needed to describe nuclei near $N = Z$. In Section 3, we briefly describe the model that we use to investigate these competing pairing effects and then in Section 4 we report some selected results for several $N = Z$ nuclei in the $2p1f$ shell. Finally, in Section 5, we summarize the key conclusions that have emerged.

2. A brief historical overview

Pairing in nuclei derives from the short-range attraction between nucleons. Pairs of nucleons can get close together to optimize the strong short-range attractive force by rotating opposite one another, achieving net orbital angular momentum zero.

Pairing is traditionally described using Bardeen-Cooper-Schrieffer (BCS) theory [2], in terms of a condensate of collective Cooper pairs. Because of the small number of active nucleons that can build pairing correlations, it is important to restore number conservation [3], which can be done using number-projected BCS theory [4].

Pairing correlations can also be combined with mean-field correlations, when necessary, through the use of Hartree Fock Bogolyubov (HFB) theory [5]. Now, however, angular momentum conservation is also lost and must be restored either exactly or approximately.

Since nucleons occur in two types (n and p), they can form four distinct types of correlated (Cooper) pairs (nn, pp, or pn), each of which can in principle have net orbital angular momentum zero and thus exploit pairing correlations. In finite nuclei, however, the neutrons and protons must be in the same major shell if they are to exploit pn pairing. Otherwise, only nn and pp pairing is important, as in typical nuclei with a neutron excess.

When pn pairing is needed, i.e. when there is not a sizable neutron excess, BCS and HFB theory can be generalized to treat all pairing modes on an equal footing. Results based on the generalized HFB theory indicate that isoscalar np pairing correlations are most important when $N \approx Z$ and when there are a reasonable number of valence neutrons and protons, precisely the regime in which quadrupole correlations are known to be important and where deformation is often present. Thus, it is important to include deformation in any analysis aimed at probing how the different kinds of pairing correlations compete in nuclear systems.

Nevertheless, there have been extensive studies of the different modes of pairing in model systems in which deformation is not present. In particular, the $SO(8)$ model [6], in which only $L = 0$ pairs contribute, has been used to obtain valuable information on the interplay of the different pairing modes. Despite the fact that the model does not contain deformation, except implicitly if the single-particle levels are not spherical, they nevertheless provide useful qualitative information. Such models, for example, make clear that generalized BCS (and thus HFB) theories cannot adequately treat the competing modes of pairing without restoration of symmetries, e.g. particle number, angular momentum *and* isospin [7].

For these reasons, we have chosen to study how the various modes of pairing compete in nuclear systems near $N = Z$ in the framework of the nuclear shell model, where it is possible to treat all pairing modes on an equal footing, preserve all symmetries and include the effects of deformation.

3. Our model

Our model consists of neutrons and protons restricted to the orbitals of the $2p1f$ shell outside a doubly-magic ^{40}Ca core and interacting via a schematic hamiltonian

$$H = \chi \left(Q \cdot Q + aP^{\dagger} \cdot P + bS^{\dagger} \cdot S + \alpha \sum_i \vec{l}_i \cdot \vec{s}_i \right) . \tag{1}$$

Here Q is the mass quadrupole operator, $P^{\dagger}$ creates a correlated $L = 0$, $S = 1$, $T = 0$ pair, $S^{\dagger}$ creates a correlated $L = 0$, $S = 0$, $T = 1$ pair, and the last term is the one-body part of a spin-orbit force.

We carry out calculations as a function of the various strength parameters and for various nuclei. We start with pure $SU(3)$ rotational motion [8] associated with the $Q \cdot Q$ interaction and then ramp up the various SU(3)-breaking terms to assess how they affect the rotational properties.

4. Calculations

4.1. Optimal hamiltonian

It is interesting to first ask whether the hamiltonian (1) is capable of meaningfully describing the properties of nuclei in this region. With the parameters $\chi = -0.05\ MeV$, $a = b = 12$, and $\alpha = 20$, we find an acceptable reproduction of all the spectra we have considered. This

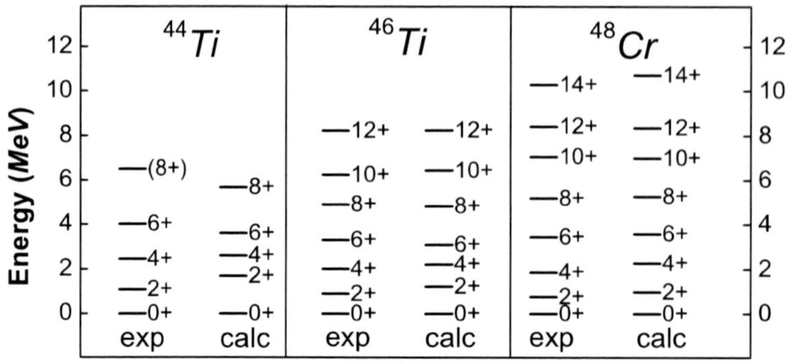

Figure 1. Comparison of experimental spectra for ^{44}Ti, ^{46}Ti and ^{48}Cr with the calculated spectra obtained using the *optimal* hamiltonian described in the text. All energies are in MeV.

is illustrated in figure 1 for ^{44}Ti, ^{46}Ti and ^{48}Cr. The calculations reproduce the known non-rotational character of ^{44}Ti and the highly rotational character known for ^{48}Cr, including its observed backbend. We refer to $a = b$ as $SU(4)$ pairing, from the dynamical symmetry that arises with this choice of parameters in the $SO(8)$ pairing model [7].

4.2. ^{44}Ti

We next focus on ^{44}Ti, which has two active neutrons and two active protons outside the inert ^{40}Ca core. In figure 2, we show the calculated energy splittings $E_I - E_{I-2}$ associated with the ground (YRAST) band as a function of the strength parameters a and b of isoscalar and isovector pairing, respectively. In these calculations we assume the optimal quadrupole strength of $\chi = -0.05\ MeV$, *but no spin-orbit interaction.* What we see is that the isoscalar and isovector pairing interactions have precisely the same effect on the properties of the ground-state rotational band, in the absence of a spin-orbit interaction. This same conclusion was known from earlier works, and is confirmed here.

Next we show in figure 3 the corresponding results *with the optimal spin-orbit term included.* Now the symmetry between isoscalar and isovector pairing is broken, even though ^{44}Ti has $N = Z$, and isovector pairing dominates.

4.3. ^{46}V

Next we turn to ^{46}V which has one additional neutron and one additional proton present. In figure 4, we show how the symmetry between isocalar and isovector pairing in the absence of a spin-orbit force is reflected in this odd-odd $N = Z$ system. In the absence of isoscalar and isovector pairing, the $J = 1^+$ state and the $J = 0^+$ state form a degenerate ground state doublet. When only isoscalar pairing is turned on (panel a), the $J = 1^+$ state is pushed down below the $J = 0^+$ state. When only isovector pairing is turned on (panel b) the reverse happens and the $J = 0^+$ is pushed down and becomes the ground state. In the $SU(4)$ limit (panel c) with equal isovector and isocalar pairing strengths, the degeneracy reappears.

In figure 5, we show what happens *in the presence of the physical spin-orbit interaction*, for equal isovector and isoscalar pairing. Now the degeneracy is broken and the 0^+ state emerges as the ground state, as in experiment. The experimental splitting is 1.23 MeV, whereas our optimal hamiltonian produces a slightly smaller splitting of 1.05 MeV.

Journal of Physics: Conference Series **387** (2012) 012018 doi:10.1088/1742-6596/387/1/012018

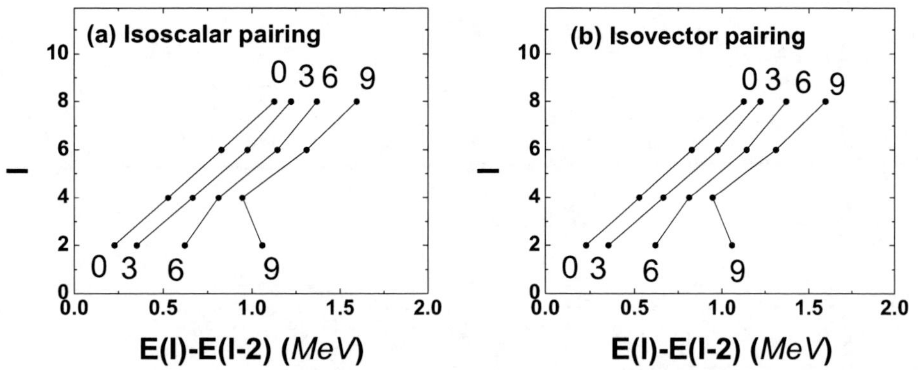

Figure 2. Spectra of the ground band of ^{44}Ti as a function of the strength of (a) the isoscalar pairing interaction and (b) the isovector pairing interaction, with no spin-orbit term present.

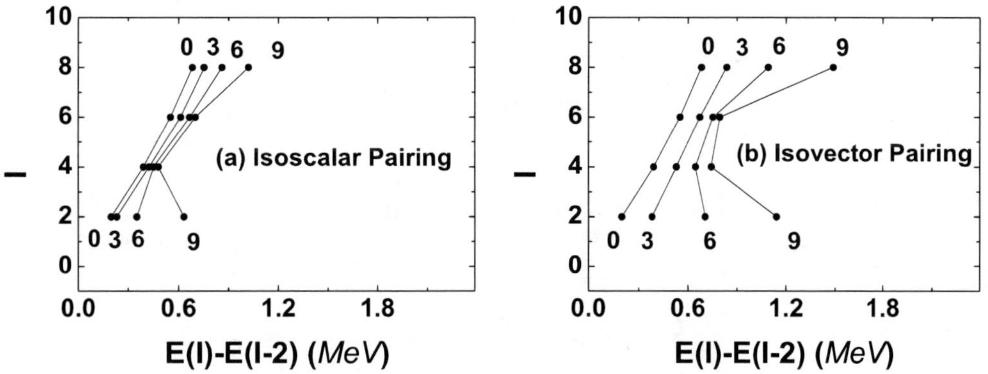

Figure 3. Spectra of the ground band of ^{44}Ti as a function of the strength of (a) isoscalar pairing interaction and (b) the isovector pairing interaction, with the optimal spin-orbit term present.

4.4. ^{48}Cr

Lastly, we consider ^{48}Cr, which again has $N = Z$, but now with two quartet-like structures present. Here we assume as our starting point both the optimal quadrupole-quadrupole force and one-body spin-orbit force and then ramp up the two pairing strengths from zero to their optimal values. The results are illustrated in figure 6, for scenarios in which we separately include isoscalar pairing, isovector pairing and $SU(4)$ pairing with equal strengths. The experimental spectrum for ^{48}Cr shows a backbend near $I = 12$, which is reproduced by our *optimal* hamiltonian. The results of figure 6 make clear that the backbend cannot be reproduced with pure isoscalar pairing, but requires isovector pairing as well.

The backbend in ^{48}Cr was discussed earlier in the context of a shell-model study with a fully

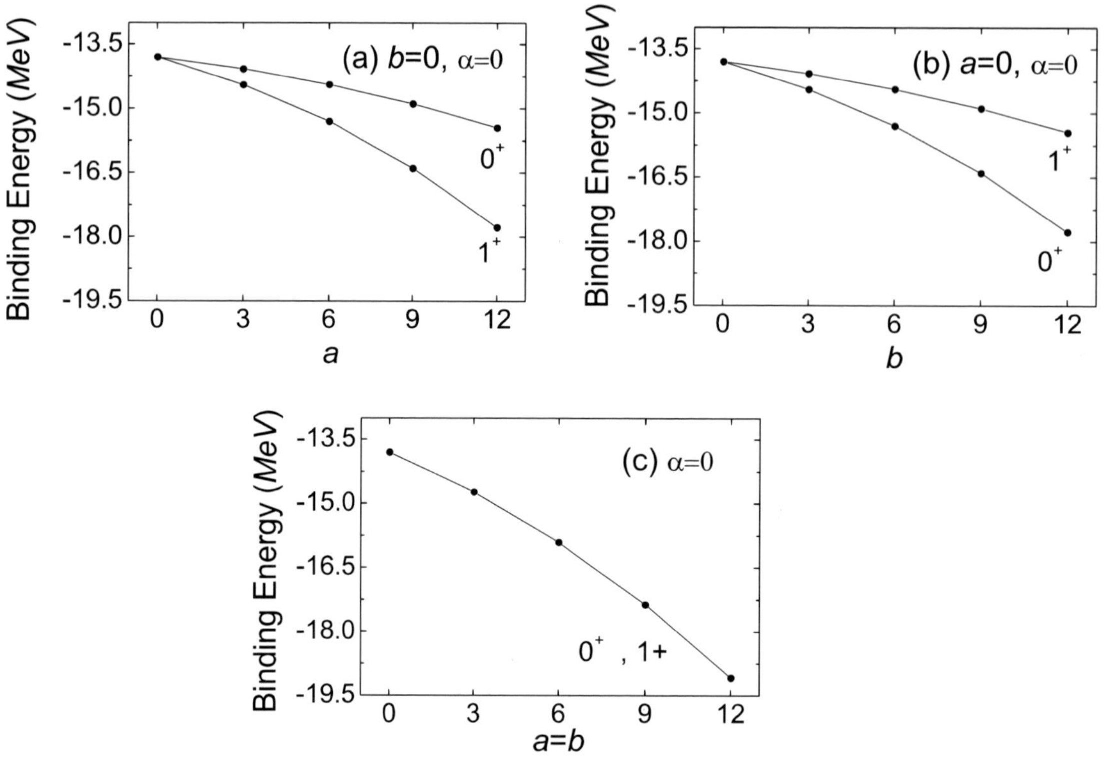

Figure 4. Calculated energies of the lowest $J^{\pi} = 0^{+}$ and 1^{+} states of ^{46}V with no spin-orbit term present, for (a) pure isoscalar pairing, (b) pure isovector pairing and (c) SU(4) pairing.

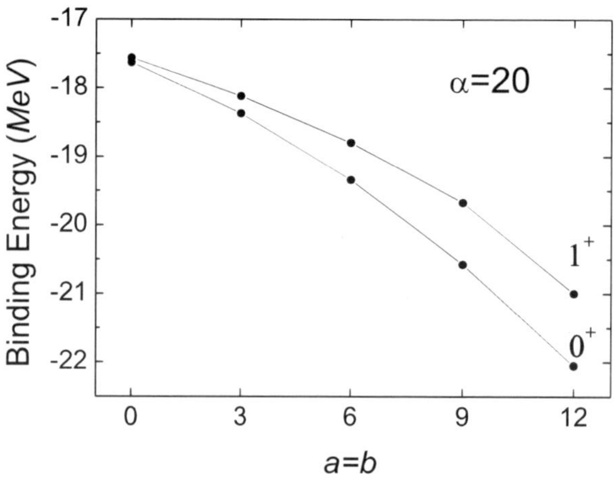

Figure 5. Calculated energies of the lowest $J^{\pi} = 0^{+}$ and $J^{\pi} = 1^{+}$ states of ^{46}V as a function of the equal strengths of isoscalar and isovector pairing, with the optimal spin-orbit term present.

Journal of Physics: Conference Series **387** (2012) 012018 doi:10.1088/1742-6596/387/1/012018

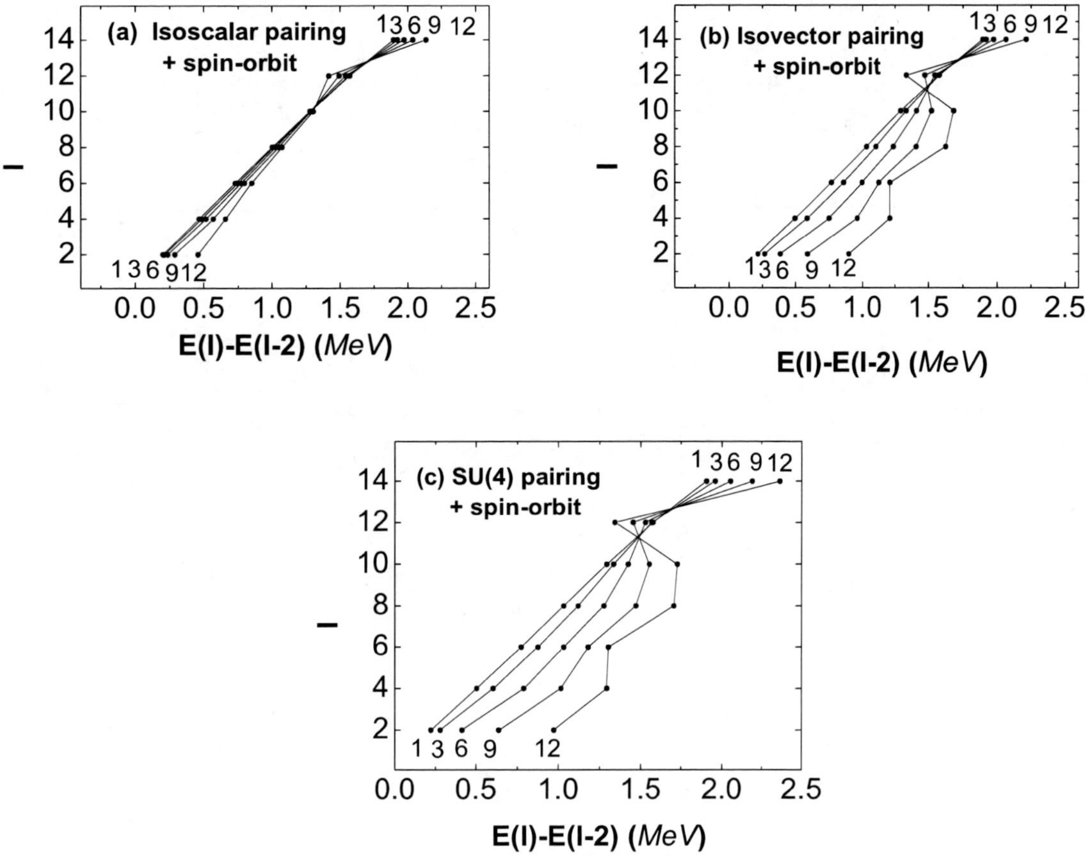

Figure 6. Calculated splittings in the ^{48}Cr ground band, for isovector, isoscalar, and SU(4) pairing, respectively, as described in the text.

realistic hamiltonian in [9], where it was first shown to derive from isovector pairing. Our results are in agreement with that conclusion. To see these points more clearly, we show in figure 7 the numbers of isovector $S^{\dagger}$ and isoscalar $P^{\dagger}$ pairs as a function of angular momentum for the optimal hamiltonian. The pair numbers are obtained by evaluating $\langle S^{\dagger} \cdot S \rangle$ and $\langle P^{\dagger} \cdot P \rangle$ and scaling them with respect to the results that would derive from pure $T = 1$ and $T = 0$ pairing hamiltonians, respectively. As in ref. [9], the contribution of isovector pairing in the $J = 0^{+}$ ground state is much larger than the contribution of isoscalar pairing. As the system cranks to higher angular momenta, the isovector pairing contribution falls off with angular momentum very rapidly eventually arriving at a magnitude roughly comparable with the isoscalar pairing contribution at roughly $J^{\pi} = 10^{+}$. As the angular momentum increases even further we see a fairly substantial increase in the isovector pairing contribution at $J^{\pi} = 12^{+}$, which according to figure 6 is where the backbend becomes prominent. After the backbend, both isoscalar and isovector pairing contributions decrease to near zero as alignment is achieved.

Since we have information not just on the states in the YRAST band but also on higher states as well, we can readily address whether the origin of the backbend that occurs in ^{48}Cr in our calculations derives from band or level crossing. In figure 8, we compare the energies of states in the YRAST band with those of the lowest excited YRARE band, a $K = 2^{+}$ band with similar intrinsic structure to the YRAST band. For the sake of comparison, we only show the

Journal of Physics: Conference Series **387** (2012) 012018 doi:10.1088/1742-6596/387/1/012018

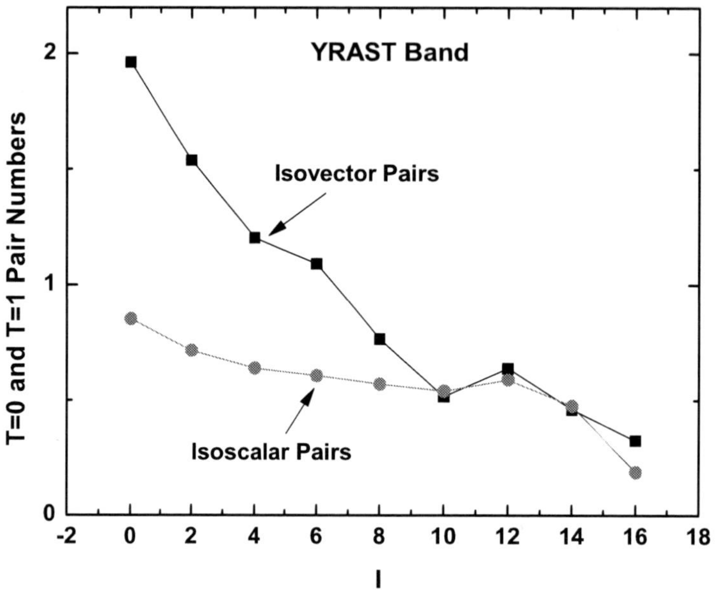

Figure 7. Calculated numbers of isovector $S^\dagger$ pairs and isoscalar $P^\dagger$ pairs in the ground (YRAST) band of ^{48}Cr for the optimal values of the hamiltonian parameters.

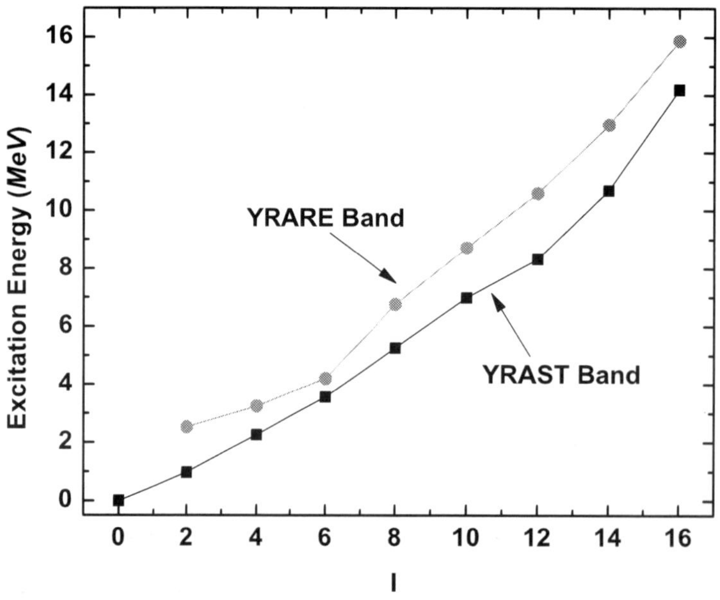

Figure 8. Calculated excitation energies of the ground (YRAST) band and the first excited (YRARE) band in ^{48}Cr for the optimal values of the hamiltonian parameters.

Journal of Physics: Conference Series **387** (2012) 012018 doi:10.1088/1742-6596/387/1/012018

even angular momentum states in the YRARE band. As can be seen from the figure, the two bands get fairly close in energy near $J = 6$, which is well below the backbend. In the vicinity of the backbend, there is no evidence of the bands getting closer together in energy. On the basis of these results, we see no evidence that the backbend in the ground band of ^{48}Cr derives from level crossing.

A possible hint as to where it might come from can be gleaned from a study of the pairing content of the two bands in the vicinity of the backbend. We already showed this information for the YRAST band in figure 7; in figure 9 we show the analogous information for the corresponding states in the YRARE band. In the YRARE band, as in the YRAST band, there is an increase in the number of isovector pairs in the vicinity of the backbend. In contrast, the number of isoscalar pairs does not change as dramatically in the YRARE band as in the YRAST band. Perhaps these results are suggesting that the backbend is driven by pairing, in some still to be defined way. We believe that these are interesting results, worthy of further study.

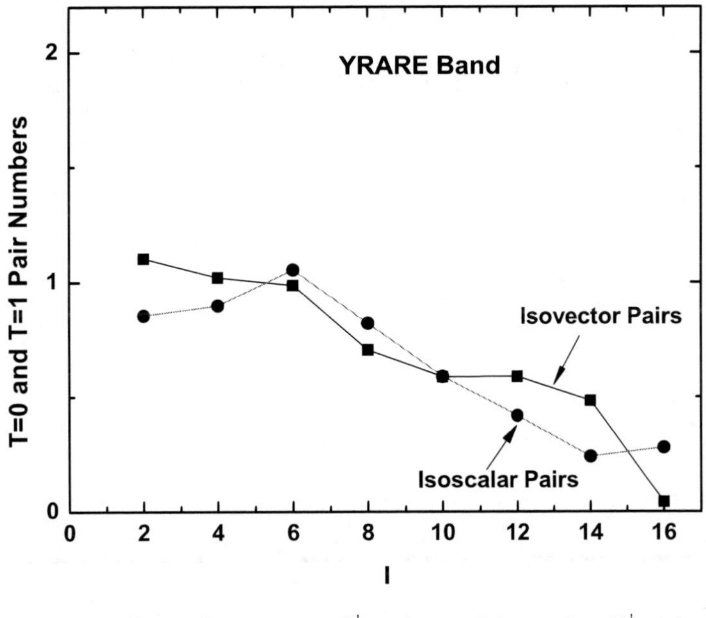

Figure 9. Calculated numbers of isovector $S^\dagger$ pairs and isoscalar $P^\dagger$ pairs in the first excited (YRARE) band of ^{48}Cr for the optimal values of the hamiltonian parameters.

5. Summary

We have reported a shell-model study of proton-neutron pairing in $2p1f$ shell nuclei using a parametrized hamiltonian that includes deformation, spin-orbit effects and isoscalar and isovector pairing and is able to describe the evolution of nuclear structure in this region. Working in a shell-model framework, we were able to assess the role of the various modes of pn pairing in the presence of nuclear deformation without violating symmetries.

Some of the key conclusions that emerged are: (1) the symmetry between isoscalar and isovector pairing effects disappears already at $N = Z$ in the presence of a spin-orbit force and isovector pairing dominates, (2) the fact that ^{46}V has a 0^+ ground state derives from the spin-orbit interaction and its relative effect on isoscalar and isovector pairing, (3) isovector pairing dominates in ^{48}Cr and produces its backbend, (4) the backbend in ^{48}Cr does not derive from the crossing of levels, and (5) isoscalar and isovector pairing exhibit a highly unusual behavior in the vicinity of the ^{48}Cr backbend.

Acknowledgments

The work of S.P., B.T. and Y.L. was supported in part by the National Science Foundation under grant # PHY-0854873, that of N.S. by the Romanian Ministry of Education and Research through CNCSIS grant Idei nr 57, that of A.P. by the projects FPA2009-13377 MICINN(Spain) and HEPHACOS S2009/ESP-1473 Comunidad de Madrid(Spain), and that of Y.M.Z. and Y.L. by the National Science Foundation of China under grant # 10975096 and by the Chinese Major State Basic Research Developing Program under grant # 2007CB815000.

References

[1] Lei Y, Pittel S, Sandulescu N, Poves A, Thakur B, Zhao Y M 2011 *Phys Rev C* **84** 044318
[2] Bardeen J, Cooper L N and Schrieffer J R 1957 *Phys Rev* **108** 1175
[3] Bohr A, Mottelson B R and Pines D 1958 *Phys Rev* **110** 936
[4] Dietrich K, Mang H J and Pradal J H 1964 *Phys Rev* **135** B22
[5] See e.g. Goodman A L 1979 *Adv Nucl Phys* **11** 263
[6] Pang S C 1965 *Nucl Phys A* **128** 497; Evans J A, Dussel G G, Macqueda E E and Perazzo R P J 1981 *Nucl Phys A* **367** 77; Dussel G G, Macqueda E E, Perazzo R P J and Evans J A 1986 *Nucl Phys A* **450** 164
[7] Dobeš J and Pittel S 1998 *Phys Rev C* **57** 688
[8] Elliott J P 1958 *Proc Roy Soc (London) A* **245** 128, 562
[9] Poves A and Martinez-Pinedo G 1998 *Phys Lett B* **430** 203

Journal of Physics: Conference Series **387** (2012) 012019 doi:10.1088/1742-6596/387/1/012019

Phase transitions for excited states in $^{16}O+\alpha \rightarrow ^{20}Ne$ within the SACM

G E Morales-Hernández[1], H Yépez-Martínez[2], P O Hess[1]

[1] Instituto de Ciencias Nucleares, UNAM, Circuito Exterior, C.U., A.P. 70-543, 04510 México, D.F., Mexico

[2]Universidad Autónoma de la Ciudad de México, Prolongación San Isidro 151, Col. San Lorenzo Tezonco, Del. Iztapalapa, 09790 México D.F., Mexico

E-mail: elgemh@hotmail.com, huitzilin_yepez@yahoo.com.mx, hess@nucleares.unam.mx

Abstract. We present the cranking of the *Semimicroscopic Algebraic Cluster Model* (SACM) for two spherical clusters. A geometrical mapping is applied and a discussion on phase transition as a function of the cranking parameter is given. This parameter can be related to the average angular momentum of the nucleus. The particular cluster system considered is $^{16}O+\alpha \rightarrow {}^{20}Ne$. We also investigate the phase transition property when the Pauli exclusion principle is not observed. We show that phase transitions may occur within the same dynamical symmetry limit.

1. Introduction

Cluster models play an important role in describing structural properties of nuclei at low energy. They contributed in the understanding of decay properties, like the α-particle decay or the emission of larger particles [1] up to the fission [2, 3]. In the usual approach, one has to antisymmetrize a two cluster system explicitly, which leads to great practical difficulties. An alternative approach was presented in [4, 5] (the above mentioned SACM), where the Pauli-principle was taken into account by the construction of a model space, which has the same content as the shell model space. While this space is microscopic, the Hamiltonian is phenomenological and contains interactions which are related to the generators of a unitary or orthogonal groups. The model was applied to many different cluster systems with great success. Also some decay properties could be understood through this model [6].

In order to study the geometrical structure, the mapping to geometrical coordinates is given in [7]. The method is based on the use of coherent states [8, 9, 10]. It is general and can be applied to any algebraic model which involves bosons. For more recent literature and excellent reviews, please consult [11, 12, 13, 14]. These considerations helped to understand the properties of a cluster system under phase transitions [15, 16]. In [15, 16] it was shown that only three effective parameters appear in the SACM for the description of phase transitions, while only two effective parameters appear, when Pauli principle is not taken into account. A model which does not observe the Pauli exclusion principle, we denote as a *Phenomenological Algebraic Cluster Model* (PACM). As a further result we showed that the PACM can, in principle, reproduce the results of the SACM when highly non-polynomial terms are included, making it more involved to apply [15].

The present contribution is related to the question, if a phase transition can appear when the total system is rotated. Phase transitions may occur when the angular momentum is increased and the phenomena may be related to the concept of *backbending* [17] or other phenomena, like the change of deformation of the united nucleus. The method in question is called *cranking* [17]. Cranking applied to cluster systems can be found in [18]. It was first applied to an algebraic model, though not a cluster model, in [19, 20]. A more recent application to an algebraic model can be found in [21]. In [22] the cranking formalism was for the first time applied to the SACM. Also the system ^{16}O+α $\rightarrow$ ^{20}Ne was considered. The reason for choosing this system is that no additional contribution from the individual clusters enter and all structure comes from the relative motion of the clusters only. However, the general structure of phase transitions was not considered and the discussion was restricted from the beginning to the $SU(3)$ dynamical symmetry.

In this contribution we will retake the question of how to crank the SACM. We again restrict to two spherical clusters, the system ^{16}O+α $\rightarrow$ ^{20}Ne, but the general treatment will already become obvious. An overview of the formalism, for both SACM and PACM, will be presented. No details will be shown and we refer to a later publication. We finally compare the results for both models and show that in the SACM a second order phase transition occurs when the angular momentum reaches a critical value, while in the PACM no phase transition occurs. We show that a phase transition in the SACM may occur within the same dynamical symmetry limit.

2. The Model and the Geometric Mapping

The basic degrees of freedom are the relative oscillations, described by the $\boldsymbol{\pi}^\dagger_m$ creation operators and by the corresponding annihilation operators $\boldsymbol{\pi}^m$ ($m = \pm 1, 0$). The products of a creation with an annihilation operator form the $U_R(3)$ group, where R refers to the relative motion. A cutoff is introduced by adding a scalar boson, whose creation and annihilation operator are $\sigma^\dagger$ and σ respectively. These bosons do not have a physical meaning but are merely introduced to create a cut-off. $N = n_\pi + n_\sigma$. The products of all types of boson creation with boson annihilation operators form a $U_R(4)$ group. For a detailed description of the SACM, please consult [4, 5].

The model space is constructed by first calculating the product $(\lambda_1, \mu_1) \otimes (\lambda_2, \mu_2) \otimes (n_\pi, 0)$, where the (λ_k, μ_k) $(k = 1, 2)$ describe the structure of the clusters 1 and 2 and n_π is the relative oscillation number. This results in a sum of many $SU(3)$ irreducible representations (irreps) (λ, μ) with a multiplicity $m_{(\lambda,\mu)}$. This list has to be compared with the content of the shell model, which can be easily obtained by computer routines, which are available to us and can be obtained on request. When a $SU(3)$ irrep in the former mentioned list does not appear in the shell model, this irrep is excluded, while when it appears it is included into the model space of the SACM. In this way, the model space of the SACM is microscopic. For the case of two spherical clusters, this procedure is equivalent to the Wildermuth condition [23], where the number of π bosons is restricted from below, in order to satisfy the Pauli exclusion principle.

The Hamiltonians, however, is phenomenologic and is written as a sum of Casimir operators of subgroups of the $U_R(4)$. When we restrict to spherical clusters, i.e. $(\lambda_k, \mu_k) = (0, 0)$, there is no contributions from the individual clusters. In $SU(3)$ notation, they are in a scalar $(0, 0)$ irrep of $SU(3)$. There are two dynamical symmetry group chains of importance, called the $SU_R(3)$ and the $SO_R(4)$ dynamical symmetries. For the case discussed here, the Hamiltonian has the form

$$\mathbf{H} = x\mathbf{H}_{SU(3)} + (1 - x)\mathbf{H}_{SO(4)} - \Omega \boldsymbol{H}_{\text{crank}} \tag{1}$$

with

$$
\begin{aligned}
\boldsymbol{H}_{SU(3)} &= \hbar\omega \boldsymbol{n}_\pi + (a - b\Delta \boldsymbol{n}_\pi)\boldsymbol{C}_2(\boldsymbol{n}_\pi, 0) + \xi \boldsymbol{L}^2 \\
\boldsymbol{H}_{SO(4)} &= \frac{c}{4}[(\boldsymbol{\pi}^\dagger \cdot \boldsymbol{\pi}^\dagger) - (\boldsymbol{\sigma}^\dagger)^2][(\pi \cdot \boldsymbol{\pi}) - (\boldsymbol{\sigma})^2] + \xi \boldsymbol{L}^2 \\
\boldsymbol{H}_{\text{crank}} &= \boldsymbol{L}_{\text{x}} \quad ,
\end{aligned} \tag{2}
$$

where the $\Delta\hat{\mathbf{n}}_\pi = \hat{\mathbf{n}}_\pi - n_0$ is the excitation number of quanta and n_0 is the minimal number of relative oscillation quanta [23], a necessary condition to satisfy the Pauli exclusion principle. The $\boldsymbol{L}_{\text{x}}$ is the x-component of the angular momentum operator and Ω is the frequency parameter which fixes the amount of the angular momentum. The structure of the Hamiltonian is equal to introduce a Lagrange multiplier [17].

The semi-classical potential is obtained by first defining a coherent state

$$
|\alpha\rangle = \frac{N!}{(N+n_o)!}\mathcal{N}_{N,n_0} \frac{d^{n_0}}{d\gamma^{n_0}} \left[\boldsymbol{\sigma}^\dagger + (\boldsymbol{\alpha}^* \cdot \boldsymbol{\pi}^\dagger)\right]^{N+n_0}|0\rangle \quad , \tag{3}
$$

where

$$
\mathcal{N}_{N,n_0}^{-2} = \frac{(N!)^2}{(N+n_0)!} \frac{d^{n_0}}{d\gamma_1^{n_0}} \frac{d^{n_0}}{d\gamma_2^{n_0}} \left[1 + \gamma_1\gamma_2(\alpha^* \cdot \alpha)\right]^{N+n_0} \tag{4}
$$

is the normalization constant. The α_m ($m = \pm 1, 0$) are the coherent state parameters. The γ is set to zero after having applied the derivatives. The corresponding coherent state for the PACM is obtained by setting $n_0 = 0$, leading to a more known form of the coherent state in algebraic models. For notational reasons, the total number of quanta is now $N + n_0$, i.e., $n_\pi + n_\sigma = N + n_0$.

Finally, calculating the expectation value of the total Hamiltonian, i.e., $\langle \alpha \mid \boldsymbol{H} \mid \alpha \rangle$, the semi-classical potential is obtained. When $\Omega = 0$, then the only relevant parameter is $\alpha = \alpha_0$, describing the relative distance of the two clusters [7, 15, 16]. However, in general we need all three components when cranking is included. For reason of simplicity, we use polar coordinates, i.e.

$$
\begin{aligned}
\alpha_{\pm 1} &= \frac{\alpha}{\sqrt{2}}\, e^{\pm i\phi} \sin(\theta) \\
\alpha_0 &= \alpha\, \cos(\theta) \quad .
\end{aligned} \tag{5}
$$

The next steps are: First, determining the minimum in the variables ϕ and θ. The minimization with respect to ϕ is easy with the result that the minimum is always at $\phi = 0$. The minimization with respect to θ is not so easy, it results in a fourth order equation, which can be expressed as $\sin 2\theta = F(\alpha)$, which is a function of α, N, n_0 and all the interaction parameters. In a second step one has to substitute the value of θ into the potential giving an effective potential V_{eff} depending only on α. Results get much easier, when only the $SU(3)$ dynamical symmetry limit is considered [22].

The Ω parameter is determined by the following steps: The expectation value of $\boldsymbol{L}_x$ is set to L, the angular momentum, which here is rather an average value. The expectation value of $\boldsymbol{L}_x$, within the SACM, is given by

$$
\langle \mathbf{L}_x \rangle = (N + n_0)\ \cos\phi\ \sin 2\theta\ \alpha^2\ \frac{F_{22}(\alpha)}{F_{00}(\alpha)} \quad , \tag{6}
$$

where $F_{pq}(\alpha)$ are polynomial functions in α [15]. The expressions get particular simple for small α. For small α, the minimum of the potential is given by

$$
\phi = 0\ , \qquad \sin 2\theta = \frac{\Omega}{4\xi n_0} \quad . \tag{7}
$$

For large α the factor $\frac{\Omega}{4\xi n_0}$ is multiplied by $[1+G(\alpha)]$, with $G(\alpha)$ being a sum of ratios of polynomials in α [15, 16]. With this, the expectation value is

$$\langle \boldsymbol{L}_x \rangle = \frac{\Omega}{4\xi} \frac{(n_0-1)}{(N+n_0-1)} = L \quad , \tag{8}$$

Which allows to deduce an approximate value for Ω, for a given L. In general, there will be a dependence on α, where one has to substitute for α the value at the minimum of the potential.

In order to study the phase transition properties, one first has to determine the minima of the potential. It turns out that $\bar{\alpha}_1 = 0$ is always an extremum, which can either be a minimum or maximum. We call this a *spherical solution*, though in the SACM the distance between the two clusters has a minimal value which is proportional to $\sqrt{n_0}$ [7], i.e., even in the intrinsic ground state a deformation may exist. There is a *deformed solution* with $\bar{\alpha}_2 > 0$. This solution not always exists. In the next step we have to find points in the phase space diagram [15, 16], where the values of the potential in both minima coincide, i.e. $V(\bar{\alpha}_1) = V(\bar{\alpha}_2)$. This describes a surface in the phase space diagram. In order to decide which order the transition is, one has to calculate on both sides the first and second derivatives *with respect to the parameters*. The order of the phase transition is one when the first derivatives in both minima are not equal, while it is of second order when the first derivatives are equal but the second are not. As shown in [15, 16] there is at most a second order phase transition in the SACM and cranking the model does not change the structure. Calculating the first and second derivatives is very involved because of the complex structure of the potential. The considerations can be greatly simplified when one considers the special structure of the potential within the SACM. As shown in [15, 16], when at the point of a phase transition the spherical and the deformed minima cross at different values of α, the phase transition is of first order, while when they merge at the point of phase transition it is of second order. For later discussion, note that the $\boldsymbol{L}_x$ is a generator of the $SO(3)$ subgroup of $SU(3)$ and thus within the $SU(3)$ dynamical symmetry limit.

As shown in this section, a general treatment is feasible but involved. This is still work in progress and the full results will be published elsewhere.

3. An application: The system ^{16}O+$\alpha \rightarrow$ ^{20}Ne

The method presented is applied to the system ^{16}O+$\alpha \rightarrow$ ^{20}Ne. In this system, the minimal number π-bosons is 8 and we choose 20 as the total number of quanta $N + n_0$, i.e., $N = 12$. The interaction parameters where fitted to experimental data, with good agreement. We will not present details here but merely report the parameter values obtained, which are $\hbar\omega = 13.2$MeV, $a = -0.5$MeV, $b = -0.009$MeV and $\xi = 0.208$MeV. It is not a surprise that the $SU(3)$ dynamical symmetry fits well the spectrum, i.e. $x = 1$ [24]. This fit was applied for $\Omega = 0$. The resulting potential shows a minimum at $\alpha = 0$, raises for larger α and finally equilibrating at a constant value for very large α, which is an effect of the finite space (N is finite) used [15, 16].

In order to simplify the consideration we will consider $\alpha << 1$. This is justified because for $\Omega = 0$ the potential of the system has a spherical minimum, so increasing Ω we may reach a point when this minimum flips over and the global minimum will be a deformed one. In other words, we expect a second order phase transition, if it exists. *Note, that this implies that even when we are in a determined dynamical symmetry limit (here $SU(3)$), still phase transition may occur* ($\boldsymbol{L}_x$ is within the $SU(3)$ dynamical symmetry), rendering the association of a dynamical symmetry to a determined phase questionable.

In the limit of $\alpha << 1$ the solution of sinθ is particular easy, namely $\sin(2\theta) \approx \frac{\Omega}{4\xi n_0}$. The potential has, up to second order on α, the approximate form

$$\begin{aligned} V(\alpha) &\approx (N+n_0)(N+n_0-1)(N+n_0-2)(-b)\left[\left(A-\Omega^2 B\right)\alpha^2\right] \\ &= 61.56\left(8.0491-0.1318\Omega^2\right)\alpha^2\ [MeV] \quad . \end{aligned} \tag{9}$$

In the last line the fitted parameter values for the system ^{16}O+α were substituted. The A is positive, using the adjusted parameters and also B is positive. The last equation gives a very intuitive picture: When $\Omega^2 < A/B$, the global minimum is spherical and for $\Omega^2 > A/B$ it flips over, i.e. the new global minimum is a deformed one. The fact that the two minima (spherical and deformed) merge at the point of phase transition, we conclude that it is a second order phase transition [15, 16]. With the parameter values used, the limiting value of Ω is given by 7.81. Using (8), leads to the angular momentum of approximately $9-10$ (9.39). Of course, this does not mean that the phase transition can be seen exactly from $L = 9-10$, probably slightly above. This phase transition happens within the $SU(3)$ dynamical symmetry. However, if this is related to backbending or another phenomenon can not be determined yet.

We also considered the PACM, when the Pauli exclusion principle is not taken into account. The equation for θ changes such that sinθ is proportional to $(1+\alpha^2)/\alpha^2$. As a result a mere constant term, proportional to Ω appears, whose effect is nothing but moving the potential up and down, by changing Ω. Therefore, in the PACM no phase transition exists when the nucleus is excited towards larger angular momentum.

4. Conclusions

In this contribution we presented the general formalism on how to introduce cranking in the SACM. We restricted to the case of two spherical clusters and the particular system considered was ^{16}O+$\alpha \rightarrow {}^{20}$Ne. The extension to systems with one or two clusters deformed is direct.

We showed that a phase transition can occur in the SACM when the angular momentum increases sufficiently. However, in the PACM no phase transition occurs, which is related to not taking into account the Pauli exclusion principle. The result also demonstrates that different phases may exist within *the same* dynamical symmetry, rendering the association of a dynamical symmetry to a single phase questionable.

The results presented here are preliminary and a list of questions still remain open. In a future work we will add the internal structure of the clusters, i.e. we take into account deformed nuclei.

Acknowledgements

We acknowledge financial help from DGAPA-PAPIIT (IN103212) and from the National Research Council of Mexico (CONACyT),

References

[1] Sandulescu A, Poenaru D N and Greiner W 1980 *Sov J Part Nucl* **11** 528
[2] Blendowske R and Walliser H 1988 *Phys Rev Lett* **61** 1930
[3] Hess P O and Mişicu Ş 2003 *Phys Rev C* **68** 064303
[4] Cseh J 1992 *Phys Lett B* **281** 173
[5] Cseh J and Lévai G 1994 *Ann Phys - New York* **230** 165
[6] Cseh J, Lévai G, Algora A and Hess P O 1997 *Rev Mex Fis* **S43** 69; Algora A, Cseh J and Hess P O 1998 *J Phys G* **24** 2111; Algora A, Cseh J and Hess P O 1999 *J Phys G* **25** 775-777; Algora A, Cseh J, Hess P O, Hunyadi M 2001 *Acta Phys Hun A* **13** 145
[7] Hess P O, Lévai G and Cseh J 1996 *Phys Rev C* **54** 2345
[8] Roosmalen O S 1982 *Algebraic Description of Nuclear and Molecular Rotation-Vibration Spectra* (Ph.D. Thesis, Groningen, Netherland)
[9] Roosmalen O S and Dieperink A E L 1982 *Ann Phys - New York* **139** 198

[10] Leviatan A and Kirson M W 1988 *Ann Phys - New York* **188** 142
[11] Cejnar P and Jolie J 2009 *Prog Part Nucl Phys* **62** 210
[12] Cejnar P and Iachello F 2007 *J Phys A* **40** 581
[13] Casten R F 2009 *Prog Part Nucl Phys* **62** 183
[14] Cejnar P, Jolie J and Casten R F 2010 *Rev Mod Phys* **82** 2155
[15] Yépez-Martínez H, Fraser P, Hess P O, Lévai G 2012 *Phys Rev C* **85** 014316
[16] Fraser P, Yépez-Martínez H, Hess P O, Lévai G 2012 *Phys Rev C* **85** 014317
[17] Ring P and Schuck P 1980 *The Nuclear Many-Body Problem* (Springer, Heidelberg)
[18] Caurier E and Grammaticos B 1983 *J Phys G* **9** L125; Marsh S and Rae W D M 1986 *Phys Lett B* **180** 185
[19] Schaaser H and Brink D M 1984 *Phys Lett B* **143** 269
[20] Schaaser H and Brink D M 1986 *Nucl Phys A* **452** 1
[21] Cejnar P 2002 *Phys Rev C* **65** 044312
[22] Yépez-Martínez H, Hess P O 2008 *Rev Mex Fís* **54** (S3) 69
[23] Wildermuth K and Tang Y C1977 *A Unified Theory of the Nucleus* (Friedr. Vieweg & Sohn Verlagsgesselschaft mbH, Braunschweig)
[24] Darai J, Cseh J and Hess P O 2011 *AIP Conf Proc* **1377** 344

Journal of Physics: Conference Series **387** (2012) 012020 doi:10.1088/1742-6596/387/1/012020

Shell model description of Ge isotopes

J. G. Hirsch and P. C. Srivastava

Instituto de Ciencias Nucleares, Universidad Nacional Autónoma de México, 04510 México, D.F., México

E-mail: hirsch@nucleares.unam.mx

Abstract. A shell model study of the low energy region of the spectra in Ge isotopes for $38 \leq N \leq 50$ is presented, analyzing the excitation energies, quadrupole moments, $B(E2)$ values and occupation numbers. The theoretical results have been compared with the available experimental data. The shell model calculations have been performed employing three different effective interactions and valence spaces. We have used two effective shell model interactions, JUN45 and jj44b, for the valence space $f_{5/2}\,p\,g_{9/2}$ without truncation. To include the proton subshell $f_{7/2}$ in valence space we have employed the fpg effective interaction due to Sorlin *et al.*, with ^{48}Ca as a core and a truncation in the number of excited particles.

1. Introduction

Ge isotopes exhibit a complex structure in their low energy spectra. The first excited 0_2^+ state shows an irregular behavior as function of the neutron number, as can be seen in Fig. 1. From ^{70}Ge to ^{72}Ge the 0_2^+ level drops in energy, rising from ^{74}Ge onwards. In the semimagic isotope ^{72}Ge, which has $N = 40$, the 0_2^+ is below than the 2_1^+. It has been interpreted in terms of configuration mixing of regular and intruder structures, leading to a shape phase transition from oblate to spherical to prolate deformed [1, 2, 3, 4].

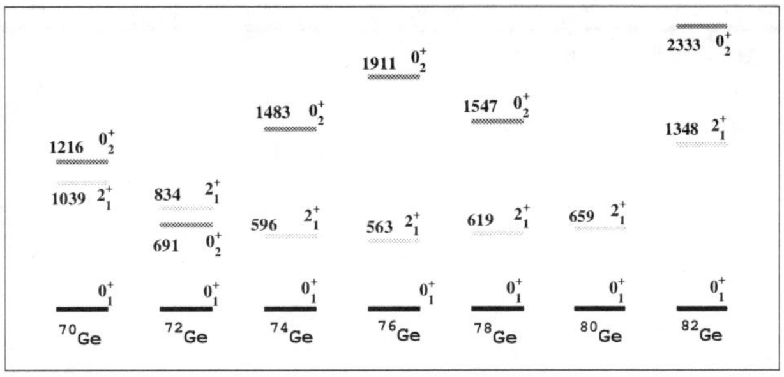

Figure 1. (Color online)Low-lying systematics of germanium isotopes.

The double beta decay of ^{76}Ge plays an upstanding role in the quest to determine the Majorana mass of the neutrino [5]. It needs a reliable description of its ground state structure, still a challenge after decades of experimental and theoretical efforts [6, 7, 8].

Journal of Physics: Conference Series **387** (2012) 012020 doi:10.1088/1742-6596/387/1/012020

In recent years there have been impressive developments in the shell model description of nuclei in the fp shell. Collective rotational bands and backbending in ^{48}Cr were described performing shel model calculations in the full fp valence shell [9]. For heavier nuclei, like the neutron-rich Cr, Mn, Fe and Co isotopes, it has been necessary to include the intruder $0g_{9/2}$ [10, 11, 12, 13, 14, 15, 16, 17]. The shell model description of the growing experimental evidence for collective behavior of Cr and Fe isotopes with $N \sim 40$ required the additional inclusion of $1d_{5/2}$ orbital [18]. While moving to heavier nuclei it was natural to exclude the $f_{7/2}$ orbital from the valence space, to understand in a shell model context recent experimental findings in Ga isotopes [19, 20, 21, 22], both the proton $f_{7/2}$ and neutron $g_{9/2}$ orbitals were required [23].

Describing the low energy spectra, electromagnetic and weak transitions in Ge isotopes employing the shell model has many difficulties. The interactions JUN45 [24] and jj44b [25] are designed to perform shell model calculations in valence space containing $f_{5/2}$, $p_{3/2}$, $p_{1/2}$, $g_{9/2}$ orbitals. They were fitted to describe the low energy spectra of many isotopes. However, a detailed shell model study of ^{70,72,74,76}Ge isotopes by Zamick's group [26] employing both JUN45 and jj44b interactions has shown that the observed $B(E2)$ transition strengths and Q values cannot be properly reproduced.

Extending Zamick's work, we have enlarged the valence space, including the $\pi f_{7/2}$ orbital to see the effect of proton excitation across Z=28, and extended the calculation up to ^{82}Ge, with $N = 50$. In the following sections we present a comparison of the low energy spectra of the Ge isotopes with $38 \leq N \leq 50$, their quadrupole moments, $B(E2)$ values and occupation numbers, obtained performing shell model calculations with the interactions JUN45 and jj44b in the valence space $f_{5/2}\,p\,g_{9/2}$ without truncation, with those done in valence fpg space with the effective interaction due to Sorlin *et al.* [27], with ^{48}Ca as a core and a truncation in the number of excited particles, and compare with the experimental information.

2. Details of calculations

The present shell model calculations have been carried out in the $f_{5/2}\,p\,g_{9/2}$ and $f\,p\,g_{9/2}$ spaces. In the $f_{5/2}\,p\,g_{9/2}$ valence space the calculations have been performed with the interactions JUN45 [24] and jj44b [25]. The single-particle energies for the $p_{3/2}$, $f_{5/2}$, $p_{1/2}$ and $g_{9/2}$ orbits employed in conjunction with the JUN45 interaction are -9.8280, -8.7087, -7.8388, and -6.2617 MeV respectively. In the case of the jj44b interaction they are -9.6566, -9.2859, -8.2695, and -5.8944 MeV, respectively. The core is ^{56}Ni, i.e. $N = Z = 28$, and the calculations are performed in this valence space without truncation.

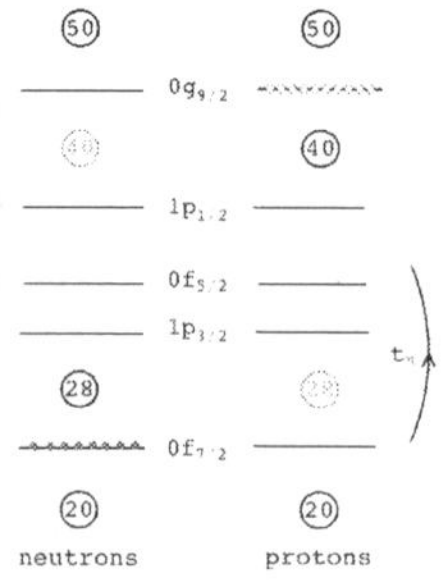

Figure 2. Model space $f\,p\,g_{9/2}$ and truncation: t_ν neutron excitations from $(p_{3/2}, f_{5/2}, p_{1/2})$ to $g_{9/2}$, t_π proton excitations from $f_{7/2}$ to $(p_{3/2}, f_{5/2}, p_{1/2})$ orbitals.

In the $f\,p\,g_{9/2}$ valence space, we use a ^{48}Ca core, i.e. only the protons are active in the

Journal of Physics: Conference Series **387** (2012) 012020 doi:10.1088/1742-6596/387/1/012020

$f_{7/2}$ orbital, and the interaction fpg reported by Sorlin *et al* [27]. The single-particle energies are 0.0, 2.0, 4.0, 6.5 and 9.0 MeV for the $f_{7/2}$, $p_{3/2}$, $p_{1/2}$, $f_{5/2}$, and $g_{9/2}$ orbits, respectively. Since the dimensionality of this valence space is prohibitively large, we have introduced a truncation by allowing t_π particle-hole excitations from the $\pi f_{7/2}$ orbital to the upper fp orbitals ($p_{3/2}$,$f_{5/2}$,$p_{1/2}$) for protons and t_ν particle-hole excitations from the upper fp orbitals to the $\nu g_{9/2}$ orbital for neutrons. This is illustrated in Fig. 2. In all the studies reported in this article the maximum allowed value for t_π and t_ν is four.

All calculations were carried out at DGCTIC-UNAM computational facility KanBalam and on Tochtli cluster computer at ICN-UNAM, using the shell model code ANTOINE [28].

3. Energy spectra

In Figs. 3 to 9 the low energy spectra of the even mass Ge isotopes ^{70,72,74,76,78,80,82}Ge are presented. In each figure the experimentally determined levels [29] are displayed on the left hand side, accompanied by the energy levels obtained through shell model calculations employing the interactions and valence spaces described above, and denoted, from left to right, JUN45, jj44b and fpg.

3.1. Fig. 3: ^{70}Ge

The first excited 2^+_1 state is observed at 1039 keV, and predicted by JUN45, jj44b and fpg interactions at 906, 737 and 478 keV, respectively. The second excited sate is the 0^+_2, predicted by JUN45 and fpg around it measured energy, but by jj44b at very high in energy. These first two excited states are reasonably described by the JUN45 interaction. The fpg interaction seems to generate a compressed spectrum for the states with angular momentum 2, 3 and 4, while the energies of the first and second excited 0^+ states are very close to the experimental ones.

3.2. Fig. 4: ^{72}Ge

As mentioned in the introduction, in ^{72}Ge the first excited state is the 0^+_2. Only the JUN45 is able to reproduce this feature. The experimental 2^+_1 at 834 keV is predicted at 811, 710 and 504 keV, respectively, by JUN45, jj44b and fpg interactions. Excluding the 0^+_2 state, the energy spectra obtained with JUN45 and jj44b are similar, while the fpg energies are again compressed for the $2^+, 3^+, 4^+$ states and the energy of the 0^+_2 state is too high.

3.3. Fig. 5: ^{74}Ge

In this isotope the first two observed excited states have angular momentum 2, followed by the two nearly degenerated states 4^+_2 and 0^+_2. There is drop in the energy of 2^+_1, 2^+_2, and 4^+_1 from ^{72}Ge to ^{74}Ge. Both JUN45 and jj44b have an acceptable description of the energies of these states. The JUN45 energies are closer to the observed ones, while the jj44b energies have the 4^+_2 and 0^+_2 in the right order. The fpg calculations fail in many ways: the energy of the 4^+_1 state is too low, and places this state below the 2^+_2, the energy of the 0^+_2 is very high.

3.4. Fig. 6: ^{76}Ge

In this Ge isotope the ordering of the first excited states is $2^+_1, 2^+_2, 4^+_1, 3^+_1, 0^+_2$. The energy spectra obtained using JUN45 and jj44b are very similar and reproduce to the observed ordering, but are displaced to higher energies by 200 keV, with the JUN45 in better agreement with the energy of the 0^+_2 state. The fpg energies are closer to the observed ones, except for the 0^+_2 state which has a higher energy (but closer to the observed one than in for the lighter Ge isotopes), and the 6^+_1 state, which is predicted below the 0^+_2, in a region where there is not reported evidence of its presence.

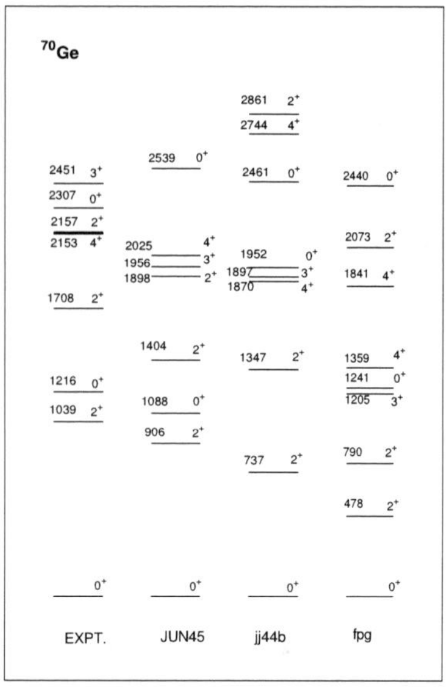

Figure 3. Calculated and experimental level schemes of ^{70}Ge, with three different interactions.

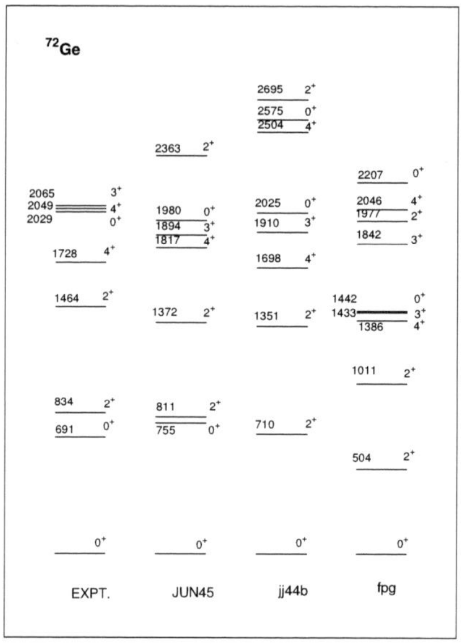

Figure 4. Calculated and experimental level schemes of ^{72}Ge, with three different interactions.

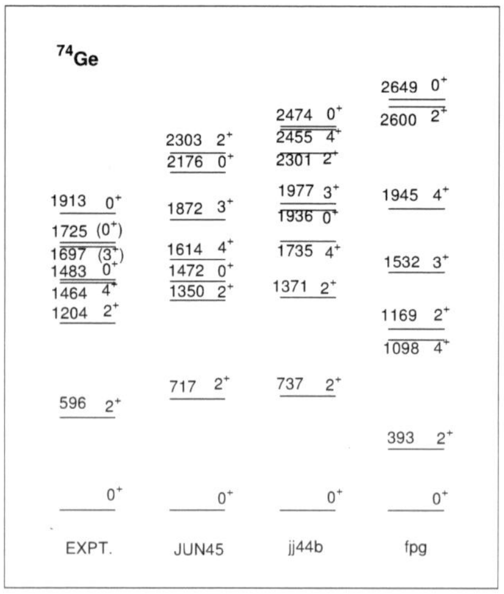

Figure 5. Calculated and experimental level schemes of ^{74}Ge, with three different interactions.

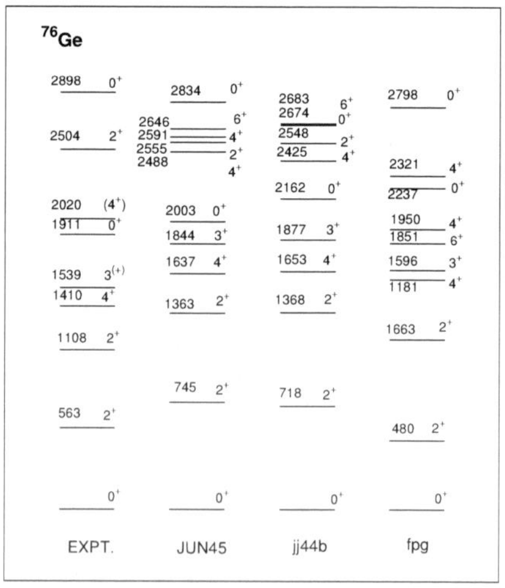

Figure 6. Calculated and experimental level schemes of ^{76}Ge, with three different interactions.

3.5. Fig. 7: ^{78}Ge
The ordering of the first excited states in ^{78}Ge is $2_1^+, 2_2^+, 0_2^+, 4_1^+$, with the states 0_2^+ and 4_2^+ very close in energy. The energy spectra obtained with *fpg* reproduce the observed energies, but displaced to a higher value by about 100 keV. The JUN45 and jj44b results follow the observed trend but the energies are 200 keV too high, and the energy of the 0_2^+ state is even higher, in particular for JUN45.

3.6. Fig. 8: ^{80}Ge
For this Ge isotope there are few well determined excited levels. The three interactions predict the 2_1^+ state around 300 keV above than experimental data.The 0_2^+ state has not yet been observed, it is predicted at 2143, 1767, 1812 keV, respectively. The experimental information seems to indicate an increase in the excitation energies of the 2_1^+, 2_2^+, and 4_1^+ states, in comparison with ^{78}Ge. The calculations with three interactions agree with this trend.

3.7. Fig. 9: ^{82}Ge
The ^{82}Ge is semimagic with N=50. The three calculations have no active neutrons, and exhibit serious difficulties to reproduce the main features of the measured levels. The energies of the 2_1^+, and 4_1^+ obtained with JUN45 are too close, the energy of the 0_2^+ state is too low for jj44b, and the excitations energies are 800 keV too high for fpg.

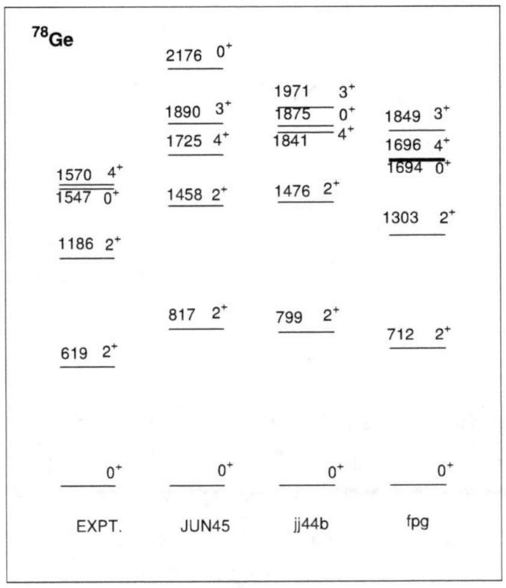

Figure 7. Calculated and experimental level schemes of ^{78}Ge, with three different interactions.

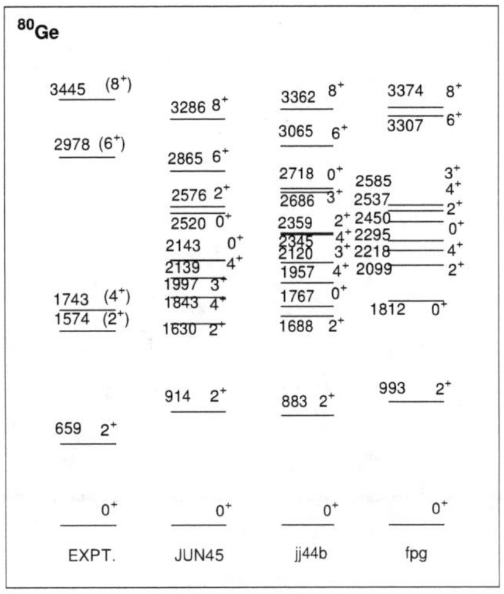

Figure 8. Calculated and experimental level schemes of ^{80}Ge, with three different interactions.

4. B(E2) Values:
The electric multipoles of order L are defined as

$$B(el, L) = \frac{1}{2J_i + 1} \mid (J_f \mid\mid \sum_i e_i r_i^L Y_L(\theta_i, \phi_i) \mid\mid \; J_i) \mid^2, \tag{1}$$

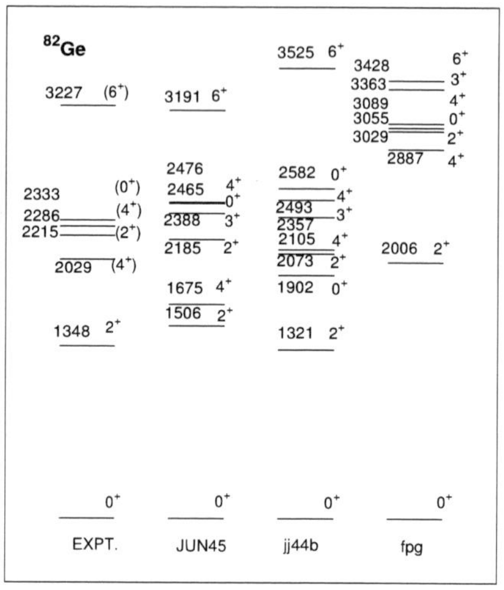

Figure 9. Calculated and experimental level schemes of ^{82}Ge, with three different interactions.

where J_i and J_f are the initial and final state spins, respectively.

The $B(E2)$ values is defined as

$$B(E2) = \frac{1}{2J_i+1} \mid (J_f \mid\mid \sum_i e_i r_i^2 Y_2(\theta_i, \phi_i) \mid\mid \; J_i) \mid^2 . \tag{2}$$

The experimentally determined $B(E2; 2_1^+ \to 0_{g.s.}^+)$ values, associated with the electromagnetic transition strength between the first excited 2^+ and the ground state, are listed in the second column of Table 1, for the even mass Ge isotopes listed in the first column. The theoretical calculations were performed employing two sets of effective charges: $e_{\text{eff}}^{\pi} = 1.5\ e$, $e_{\text{eff}}^{\nu} = 0.5\ e$ and $e_{\text{eff}}^{\pi} = 1.5\ e$, $e_{\text{eff}}^{\nu} = 1.1\ e$, displayed in the same column separated by a "/".

Except for ^{82}Ge, where in the three models the neutrons are not active and do not contribute at all to the transition, leading to small theoretical values, for all the other isotopes the models can reproduce the observed transition intensities. For JUN45 the second set of effective charges works better, jj44b would need some intermediate values for the neutron effective charge, and the large values obtained with *fpg* are probably related with the contribution of the protons in the $f_{7/2}$ orbital, which could be compensated with smaller effective charges

5. Quadrupole moments:

The electric quadrupole moment operator is defined as

$$Q_z = \sum_{i=1}^{A} Q_Z(i) = \sum_{i=1}^{A} e_i(3z_i^2 - r_i^2). \tag{3}$$

The spectroscopic quadrupole moment is defined as

$$Q_s(J) = \langle J, m = J \mid Q_2^0 \mid J, m = J\rangle = \sqrt{\frac{J(2J-1)}{(2J+1)(2J+3)(J+1)}} \langle J \mid\mid Q \mid\mid J\rangle. \tag{4}$$

Table 1. The measured [4] and calculated $B(E2)$ values, given in units of $10^{-3}e^2b^2$. Two set of effective charges are employed: $e^{\pi}_{\text{eff}} = 1.5\ e$, $e^{\nu}_{\text{eff}} = 0.5\ e$ and $e^{\pi}_{\text{eff}} = 1.5\ e$, $e^{\nu}_{\text{eff}} = 1.1\ e$, shown separated "/".

Nucleus	Expt.	JUN45	jj44b	*fpg*
^{70}Ge	36(4)	24.8/47.5	33.7/67.7	55.5/99.5
^{72}Ge	40(3)	26.2/50.8	35.3/71.2	44.3/78.1
^{74}Ge	60(3)	30.7/59.1	36.7/70.6	51.5/94.4
^{76}Ge	46(3)	31.3/56.9	34.8/63.9	53.6/93.7
^{78}Ge	44(3)	28.5/48.0	31.5/53.8	48.9/79.8
^{80}Ge	28(5)	19.2/30.1	22.9/34.8	33.3/47.5
^{82}Ge	25(5)	8.1/8.1	15.9/15.9	22.7/22.7

The $E2$ operator in Eq. (2) is expressed as a function of the spherical tensor components:

$$Q^0_2 = Q_z = \sqrt{\frac{16\pi}{5}} \sum_{i=1}^{A} e_i r_i^2 Y_{20}(\theta_i \phi_i). \tag{5}$$

In Table 2 we present the experimental and calculated quadrupole moments, obtained with the three different effective interactions with the two set of effective charges, for the Ge isotopes listed in the first column, and for the first, and in some cases the second, 2^+ state. At striking variance with the results obtained for the $B(E2)$ values, the quadrupole moments represent a challenge which surpass the capabilities the three models. The sign of the quadruple moments is related with the type of deformation: oblate of prolate. It does not change with the rescaling associated with the use of effective charge. For ^{72}Ge the sign of the predicted quadrupole moment is positive for the first 2^+ state, and negative for the second. The reported experimental signs are the opposite. For ^{74}Ge only the signs predicted using jj44b coincide with the experimental ones, but the theoretical magnitudes are three to five times smaller. Also for ^{76}Ge the signs obtained using jj44b are the same give by the experimentalists, and in this case the magnitudes also agree, while the quadrupole moments obtained with JUN45 are very small, and those given by *fpg* are large with the opposite sign. The theoretical predictions for ^{78}Ge have all negative signs, but differ up to one order of magnitude between the different models.

6. Occupation numbers

In Fig. 10, we show the proton (a) and neutron (b) occupation numbers of the different orbitals for the ground and first excited 2^+ states of the isotopes $^{70-82}$Ge. The general trends are very interesting. The proton occupancies increase smoothly in the $f_{5/2}$ orbital, at the expense of the $p_{3/2}$. It suggests that there is a crossing between these two orbitals due to the interactions with the neutrons. This crossing is predicted to take place at N=38 with jj44b, at N=40 with fpg, and at N=42 employing JUN45. In the neutron sector increasing the neutron number leads to a nearly linear grow in the occupation of the $g_{9/2}$ orbital, and with a smaller degree of the $f_{5/2}$ occupations.

In Table 3 we have tabulated the occupancies of proton and neutron orbitals for the ^{76,78}Ge isotopes, for the ground and the first excited 2^+ states. Some of these occupancies have been determined experimentally for ^{76}Ge [6, 7], and these values with their uncertainties are shown in the second row. In the case of ^{76}Ge the proton occupancies predicted with JUN45 and jj44b are in general similar, and both close to those determined experimentally. A clear consequence of opening the $\pi 0f_{7/2}$ and closing the $\pi 0g_{9/2}$ orbitals in the *fpg* calculations is the large occupation

Table 2. The measured [4] and calculated quadrupole moments, in $10^{-2}eb$. Two sets of effective charges are employed: $e^{\pi}_{\text{eff}} = 1.5\ e$, $e^{\nu}_{\text{eff}} = 0.5\ e$ and $e^{\pi}_{\text{eff}} = 1.5\ e$, $e^{\nu}_{\text{eff}} = 1.1\ e$, separated by "/".

Nucleus	I	Expt.	JUN45	jj44b	fpg
^{70}Ge	$Q(2_1^+)$	+3(6) or +9(6)	+10.00/+16.94	+15.13/+24.80	+29.55/+41.92
^{72}Ge	$Q(2_1^+)$	-12(8)	+12.83/+21.55	+10.96/+18.61	+31.28/+44.12
	$Q(2_2^+)$	+23(8)	-13.46/-22.27	-11.31/-19.02	-32.01/-44.79
^{74}Ge	$Q(2_1^+)$	-19(2)	+11.95/+19.86	-5.89/-6.35	+39.25/+54.54
	$Q(2_2^+)$	+26(6)	-11.46/-18.54	+5.39/+5.98	-39.47/-54.56
^{76}Ge	$Q(2_1^+)$	-14(4)	+1.69/+4.55	-14.51/-18.63	+30.04/+40.89
	$Q(2_2^+)$	+28(6)	+0.02/-1.94	+15.52/+20.15	-29.36/-39.75
^{78}Ge	$Q(2_1^+)$	-	-10.90/-13.00	-15.53/-19.48	-2.41/-2.21
^{80}Ge	$Q(2_1^+)$	-	-23.45/-28.84	-23.45/-29.03	-26.90/-32.11
^{82}Ge	$Q(2_1^+)$	-	-19.56/-19.56	-18.21/-18.21	-18.46/-18.46

of the $\pi 0f_{5/2}$, predicting about one proton more than the number measured in this orbital. In the neutron sector the jj44b occupations fully agree, within the experimental errors, with the observed ones. The jj44b and JUN45 occupations are quite similar, and both show a slightly larger occupation in the $\nu 0f_{5/2}$ at the expense of the $\nu 0g_{9/2}$ orbital. The proton occupations in ^{78}Ge exhibit a small displacement from the $\pi 1p_{3/2}$ and $\pi 1p_{1/2}$ orbitals to the $\pi 0f_{5/2}$ orbital, more noticeable for in the fpg scheme. The neutron occupations are similar for the three interactions.

In both isotopes the occupations calculated for the first excited 2^+ state are very close to those of the ground state. It is worth to mention that the $\nu 0g_{9/2}$ occupations are larger, by nearly to units, than those expected in the extreme shell model estimation: 4 for ^{76}Ge and 6 for ^{78}Ge.

Table 3. Occupation of proton and neutron orbitals for ^{76}Ge and ^{78}Ge isotopes in $f_{5/2}pg_{9/2}$ and $fpg_{9/2}$ spaces.

Interaction	Nucleus	I	$\pi 0f_{7/2}$	$\pi 1p_{3/2}$	$\pi 1p_{1/2}$	$\pi 0f_{5/2}$	$\pi 0g_{9/2}$	$\nu 1p_{3/2}$	$\nu 1p_{1/2}$	$\nu 0f_{5/2}$	$\nu 0g_{9/2}$
Expt.[6, 7]	^{76}Ge	g.s.		1.75±0.15		2.04±0.25	0.23±0.25	4.87±0.20		4.56±0.40	6.48±0.30
JUN45	^{76}Ge	0_1^+		1.58	0.27	1.90	0.23	3.54	1.43	5.18	5.85
		2_1^+		1.54	0.29	1.97	0.20	3.57	1.41	5.12	5.89
jj44b	^{76}Ge	0_1^+		1.44	0.34	2.00	0.22	3.62	1.48	4.74	6.16
		2_1^+		1.42	0.38	2.02	0.18	3.65	1.48	4.69	6.17
fpg	^{76}Ge	0_1^+	7.49	1.55	0.28	2.73	-	3.63	1.28	5.22	5.87
		2_1^+	7.47	1.57	0.23	2.73	-	3.61	1.28	5.20	5.91
JUN45	^{78}Ge	0_1^+		1.39	0.23	2.16	0.22	3.66	1.57	5.57	7.19
		2_1^+		1.40	0.27	2.15	0.18	3.68	1.57	5.57	7.17
jj44b	^{78}Ge	0_1^+		1.30	0.30	2.20	0.20	3.75	1.59	5.22	7.45
		2_1^+		1.28	0.35	2.20	0.16	3.76	1.59	5.19	7.46
fpg	^{78}Ge	0_1^+	7.51	0.95	0.20	3.33	-	3.60	1.24	5.58	7.57
		2_1^+	7.48	1.00	0.23	3.28	-	3.57	1.22	5.56	7.65

7. Concluding remarks

We have presented three sets of shell model calculations for even $^{70-82}$Ge isotopes in the $f_{5/2}pg_{9/2}$ and $fpg_{9/2}$ spaces, employing recently developed interactions. The performance of the three calculation schemes is mixed. The JUN45 interaction in the $f_{5/2}pg_{9/2}$ valence space allowed

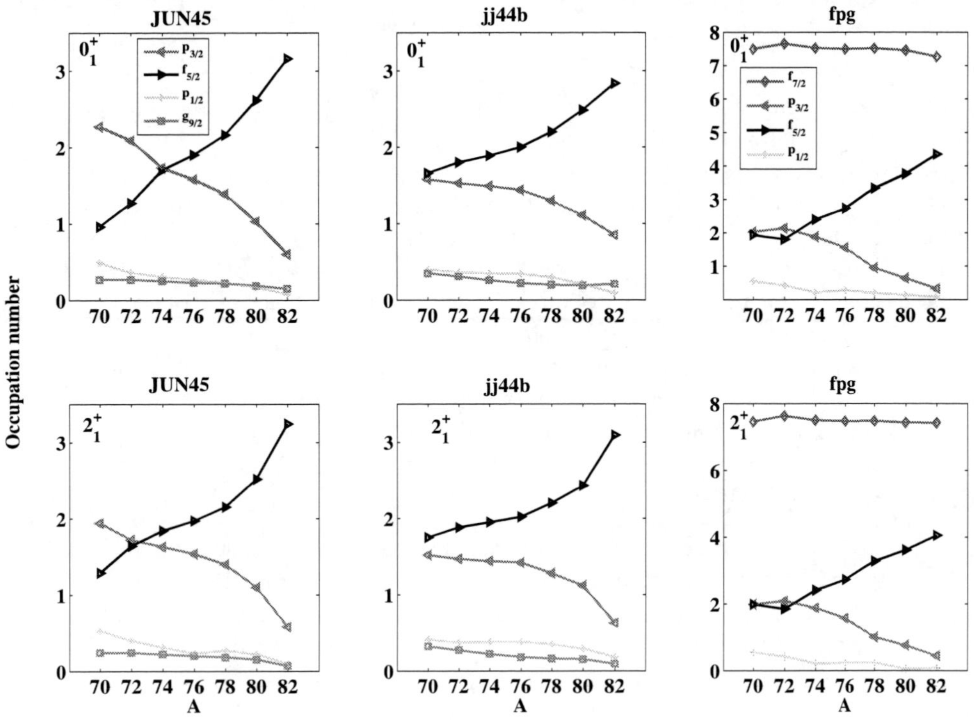

(a) Proton

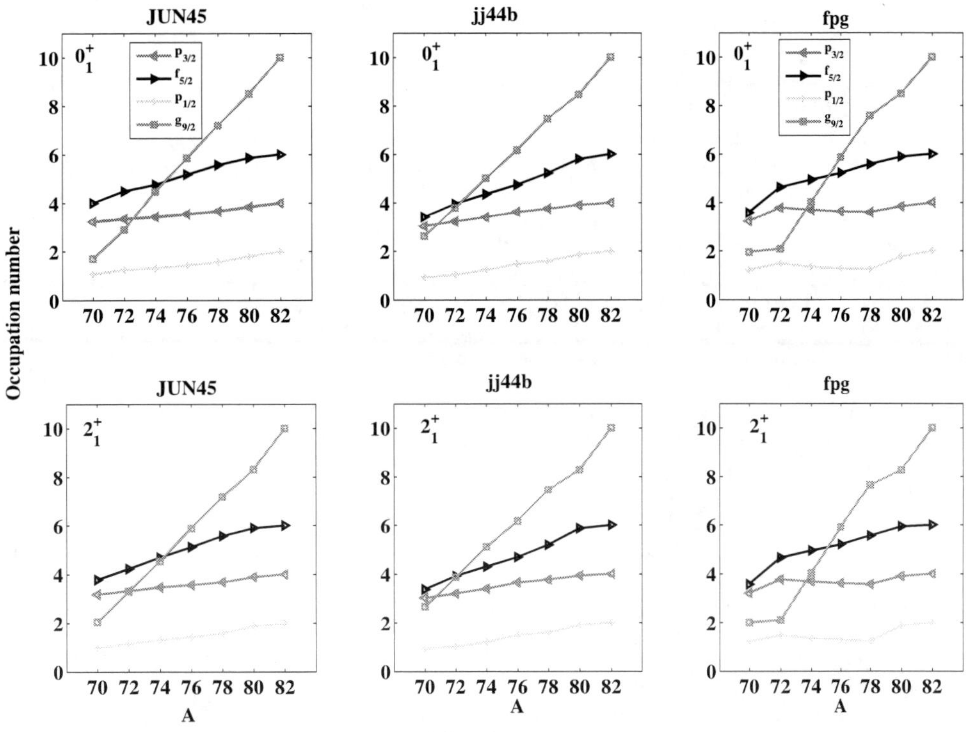

(b) Neutron

Figure 10. (Color online) Proton/Neutron occupation numbers of the JUN45 and jj44b ($p_{3/2}$, $f_{5/2}$, $p_{1/2}$ and $g_{9/2}$ -shell orbits) and fpg ($f_{7/2}$, $p_{3/2}$, $f_{5/2}$, $p_{1/2}$ -shell orbits) interactions- for two low-lying states in even-even Ge isotopes. (Upper panel) 0_1^+ states;(lower panel) 2_1^+ states.

a good description of the observed low energy spectra in ^{70,72,74}Ge and ^{80}Ge, while the fpg interaction in the $fpg_{9/2}$ space worked better for ^{76,78}Ge. The calculations for ^{82}Ge, having only the proton orbitals active, were very limited in any case. The $B(E2)$ values were reproduced without much problem in the three schemes, while the quadrupole moments represented a very difficult challenge. The jj44b interaction was able to reproduce the measured quadrupole moments and occupations in ^{76}Ge.

One obvious conclusion related to the shell model description of Ge isotopes is the need to include upper orbitals in the neutron sdg shell, following the developments introduced by Lenzi et al [18]. The theoretical description of the low energy spectra of isotopes in this mass region would require the inclusion of many orbitals, where we find the known limitation due to the enormous size of the Hilbert space, and the spurious center of mass remotion.

Acknowledgements

We would like to thank E. Padilla-Rodal and A. Poves for useful discussions during this work. All calculations were carried out using the computational facilities KanBalam at DGCTIC-UNAM and Tochtli at ICN-UNAM. P. Srivastava is a postdoctoral fellow of DGAPA-UNAM. This work was supported in part by Conacyt, México, and DGAPA, UNAM project IN103212.

References

[1] Leconte R, Irshad M, Landsberger S, Paradis P and Monaro M 1980 *Phys. Rev.* C **22** 1530
[2] Toh Y *et al.* 2002 *AIP Conf. Proc.* **610** 793
[3] Padilla-Rodal E *et al* 2005 *Phys. Rev. Lett.* **94** 122501
[4] Padilla-Rodal E 2010 *Journal of Physics: Conference Series* **239** 012014
[5] Elliott S R and Vogel P 2002 *Annu. Rev. Nucl. Part. Sci.* **52** 115; Suhonen J and Civitarese O 1998 *Phys. Rep.* **300** 123; Faessler A and Ŝimkovic F 1998 *J. Phys.* G **24** 2139
[6] Schiffer J P *et al* 2008 *Phys. Rev. Lett.* **100** 112501
[7] Kay B P *et al* 2009 *Phys. Rev.* C **79** 021301 (R)
[8] Moreno O, Moya de Guerra E, Sarriguren P, and Faessler A 2010 *Phys. Rev.* C **81**; Suhonen J 2011 *Nucl. Phys.* A **853** 36; Gómez-Cadenas J J, *et al* 2011 *Jour. Cosm. Astr. Phys.* **06** 007 and references therein
[9] Caurier E *et al.* 1995 *Phys. Rev. Lett.* **75** 2466
[10] Lurandi S. *et al.* 2007 *Phys. Rev.* C **76** 034303
[11] Valiente-Dobón J J*et al* 2008 *Phys. Rev.* C **78** 024302
[12] Kaneko K, Sun Y, Hasegawa M, and Mizusaki T 2008 *Phys. Rev.* C **78** 064312
[13] Srivastava P C and Mehrotra I 2009 *J. Phys. G: Nucl. Part. Phys.* **36** 105106
[14] Srivastava P C and Mehrotra I 2010 *Phys. At. Nucl.* **73** 1656
[15] Srivastava P C and Mehrotra I 2010 *Eur. Phys. J.* A **45** 185
[16] Sun Y, Yang Y-C, Liu H-L, Kaneko K, Hasegawa M, and Mizusaki T 2009 *Phys. Rev.* C **80** 054306
[17] Srivastava P C and Kota V K B 2011 *Phys. At. Nucl.* **74** 971
[18] Lenzi S M, Nowacki F, Poves A and Sieja K 2010 *Phys. Rev.* C **82** 054301
[19] Cheal B *et al* 2010 *Phys. Rev. Lett.* **104** 252502
[20] Mané E *et al* 2011 *Phys. Rev.* C **84** 024303
[21] Diriken J *et al* 2010 *Phys. Rev.* C **82** 064309
[22] Cheal B *et al* 2010 *Phys. Rev.* C **82** 051302
[23] Srivastava P C 2012 *J. Phys. G: Nucl. Part. Phys.* **39** 015102
[24] Honma M, Otsuka T, Mizusaki T and Hjorth-Jensen M 2009 *Phys. Rev.* C **80** 064323
[25] Brown B A and Lisetskiy A.F. (unpublished)
[26] Robinson S J Q, Zamick L, Sharon Y Y 2011 *Phys. Rev.* C **83** 027302
[27] Sorlin O *et al* 2002 *Phys. Rev. Lett.* **88** 092501
[28] Caurier E, Martínez-Pinedo G, Nowacki F, Poves A, and Zuker A P 2005 *Rev. Mod. Phys.* **77** 427
[29] http://www.nndc.bnl.gov/ensdf/

XXXV Symposium on Nuclear Physics IOP Publishing
Journal of Physics: Conference Series **387** (2012) 012021 doi:10.1088/1742-6596/387/1/012021

Quantum phase transitions in the LMG model by means of quantum information concepts

O Castaños, R López-Peña, E Nahmad-Achar, and J G Hirsch

Instituto de Ciencias Nucleares, Universidad Nacional Autónoma de México
Apartado Postal 70-543, 04510 México, D. F., México

E-mail: ocasta@nucleares.unam.mx

Abstract.
Quantum phase transitions are currently studied in several fields of physics, as examples one finds: in nuclear physics the shape phase transitions within the Interacting Boson Model; in quantum optics two level atoms interacting with a one mode electromagnetic radiation in a cavity; and in condensed matter in the analysis of the behavior of spin systems. In this contribution we employ the fidelity and its susceptibility, two concepts widely used in quantum information theory, to determine the quantum phase transitions of the Lipkin-Meshkov-Glick (LMG) model together with its scaling properties. By means of these concepts, we propose a quantum method to determine the crossings and anti-crossings present in a model Hamiltonian as a function of the control parameters of the model. A review of the separatrix of the LMG model is done to compare the results obtained by means of quantum information concepts with those related with the singular behavior of the energy surface of the model, which is the expectation value of the Hamiltonian with respect to spin coherent states.

1. Introduction

The quantum phase transitions of many body systems are characterized by a sudden change in the ground state properties. They are different to the thermal phase transitions because they are due entirely to quantum fluctuations. They are described typically as due to a non-analytic behavior of the ground state properties when the Hamiltonian system crosses a transition point. Although one of the most important manifestations of a quantum phase transition occurs in the energy spectra of the physical systems, it can be observed also in many other observables.

The crossing of levels in the spectrum of physical system is an indication of a first order transition, while the second order ones correspond to other causes and they are continuous. This latter type of transitions can be characterized by means of the spontaneous symmetry breaking formalism developed by Ginzburg, Landau and Wilson, where the order parameters play a crucial role. In particular the Ginzburg-Landau theory is related with the phase transitions of a material between the superconductor state to the normal one [1]. A manner to relate a variety of different phenomena in condensed matter theory is through the renormalization group, which originated to reconcile divergences in continuos theories with the finite experimental results. This is done by means of a cutoff procedure which in the modern vision lets us separate the modes of the physical system [2].

The study of complex systems can be simplified by analyzing their static and dynamic entanglement properties. When a physical system approximates a critical point, the structure

Journal of Physics: Conference Series **387** (2012) 012021 doi:10.1088/1742-6596/387/1/012021

of the entanglement suffers a transition. For bipartite systems characterized by pure states the best measure of entanglement is provided by the von Neumann entropy, while for physical systems described by a mixed state one uses other quantities like the formation entanglement (concurrence), and the relative entropy [3].

In the literature there are two main forms of characterizing a quantum phase transition, the one introduced by Ginzburg and Landau where the catastrophe formalism can be very important and more recently those related with the use of entanglement entropy, fidelity and fidelity susceptibility. The fidelity approach is an alternative procedure to understand the critical phenomena, that is the physics associated to the critical points in the thermodynamical sense. It includes relations between different quantities, power laws of the quantities described by critical exponents and universality [4].

2. Description of phase transitions

The classical theory of phase transitions is adjusted to the framework of elementary phase transitions, for describing a physical system as a family of potentials $V = V(x;c)$, which is dependent of n real variables or order parameters x, and k real control parameters c. It is generally assumed that the physical system is described for the value x that minimizes the potential, at least locally. Then the study of the system is reduced to the equilibrium and stability properties of the family of potentials. In general, almost for all values of the control parameters the potential will have isolated critical points. The relations $\frac{\partial V}{\partial x_k} = 0$ constitute a system of n equations that can be considered state equations and they let us get $x^p(c)$, the critical points in terms of the control parameters. The local stability of one of these points is determined by using the hessian matrix evaluated at $(x^p(c), c)$. A phase transition happens when the point x describing the ground state of the physical system jumps from a critical point to another [5].

Usually the phase transitions occur when the control parameters are varied, frequently the time is chosen as the parametric variable, and then one has $c_\alpha \to c_\alpha(t)$, $x^p(c_\alpha) \to x^p(c_\alpha(t))$ and $V^p(x^p(c_\alpha), c) \to V^p(t)$. In the catastrophe formalism a phase transition occurs when the curve $c_\alpha(t)$ crosses a component of the so called *separatrix*. It is defined as the union of the Maxwell S_M and bifurcation S_B sets which are formed by the critical points of the potential energy [6].

The determinant of the hessian matrix, evaluated at the critical points,

$$V_{i,k} = \left. \frac{\partial^2 V}{\partial x_i \partial x_k} \right|_{x^p(c)} , \tag{1}$$

allows a determination of the degenerated critical points. This region of points defines the bifurcation sets while the Maxwell sets are determined by means of the Clausius-Clapeyron equations [6]

$$V^p = V^{p+1} , \qquad \left(\frac{\partial V^p}{\partial c_\alpha} - \frac{\partial V^{p+1}}{\partial c_\alpha} \right) \delta c_\alpha = 0 . \tag{2}$$

In the delay convention the physical system remains in a stable or metastable equilibrium state until that state disappears while in the Maxwell convention the physical system is the one that globally minimizes the potential energy. The order of a phase transition is given according to the Ehrenfest classification, that is one gets a phase transition of order m if

$$\lim_{\epsilon \to 0} \left. \frac{d^j V^p(t)}{dt^j} \right|_{t=t_0+\epsilon} = \lim_{\epsilon \to 0} \left. \frac{d^j V^p(t)}{dt^j} \right|_{t=t_0-\epsilon} \tag{3}$$

is satisfied for all $j = 0, 1, 2, \cdots m-1$ but not for $j = m$. The phase transition is called local or soft if the following relation is satisfied

$$\lim_{\epsilon \to 0}(x^p(t_0+\epsilon) - x^q(t_0-\epsilon)) = 0\,, \tag{4}$$

where x^p and x^q denote the values of x at which V attains its minimum, after and before the transition, respectively; otherwise is called hard or non-local [6].

To show the procedure, one can consider the potential energy surface associated to the cusp catastrophe [5, 6], which has one order variable and two control parameters

$$V(x; a, b) = \frac{1}{4}x^4 + \frac{a}{2}x^2 + bx\,. \tag{5}$$

This potential energy describes an anharmonic oscillator. It is immediate that the critical points are determined by solving the third order algebraic equation

$$x^3 + ax + b = 0\,. \tag{6}$$

whose solutions can be given in terms of Cardan's formulas,

$$\begin{aligned} x_1 &= \left(-\frac{b}{2} + \sqrt{D}\right)^{1/3} + \left(-\frac{b}{2} - \sqrt{D}\right)^{1/3}, \\ x_2 &= \xi\left(-\frac{b}{2} + \sqrt{D}\right)^{1/3} + \xi^2\left(-\frac{b}{2} - \sqrt{D}\right)^{1/3}, \\ x_3 &= \xi^2\left(-\frac{b}{2} + \sqrt{D}\right)^{1/3} + \xi\left(-\frac{b}{2} - \sqrt{D}\right)^{1/3}, \end{aligned} \tag{7}$$

where we have defined the discriminant $D = \left(\frac{b}{2}\right)^2 + \left(\frac{a}{3}\right)^3$ and the complex parameter $\xi = \exp(-i\frac{2\pi}{3})$.

In Fig. 1 the bifurcation and Maxwell sets are plotted in the control parameter space. The bifurcations are defined when the discriminant takes the value zero, that is when there are degenerated critical points, while the Maxwell set corresponds to the case of two critical points with the same value for the potential energy surface. The latter happens for the line $b = 0$ with $a \leq 0$, to allow for the existence of the critical point. In the region of points above the fold lines (the points where the manifold of critical points is singular) of the critical points, the potential energy surface has only one minimum, deformed to the left for $b > 0$ and to the right for $b < 0$. On the bifurcation set the minima take place away from $x = 0$, for negative x values when $b < 0$ and positive x values for $b > 0$. Inside the bifurcation set the potential energy has three critical points, one maximum and two minima, and exactly at the Maxwell set the minima for positive and negative values of the order variable are of equal depth. At the points on the bifurcation the phase transitions are of order zero, the crossing through the Maxwell set yields a phase transition whose order generally it depends of the curve used to cross it. Finally by going directly along the line $b = 0$ the transition is of order two.

The quantum mechanical study of the cusp potential energy surface has been done by Gilmore et al [7]. The spectrum presents two different regimes: at low excitation energies is linear in the number of phonon excitations n, while at high excitation energies the spectrum behaves like $n^{4/3}$. The spectrum has single and double mode regions, the one-mode case occurs for $a > 0$ where the spectrum is equally spaced, while for $a < 0$ it is again equally spaced but there is a softening of the spring constant at middle energies. The bimodal case takes place inside the fold

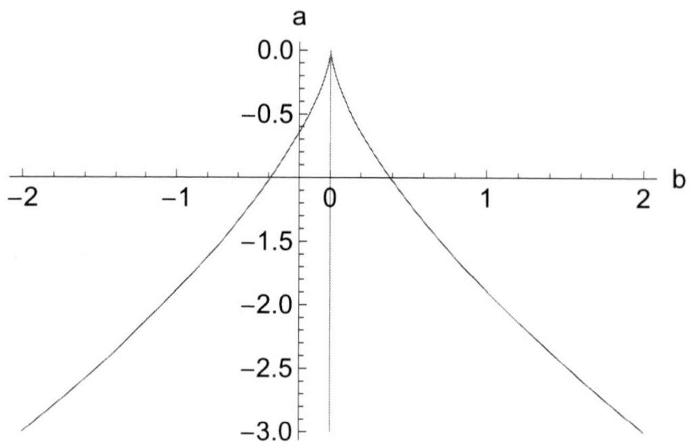

Figure 1. The separatrix for the cusp catastrophe. The blue line (cusped) corresponds to the bifurcation set while the red (vertical) line to the Maxwell set. The separatrix distinguishes physical systems with mixtures of two phases below the fold lines and a pure phase above them.

lines of the separatrix and there are two bands with a harmonic behavior at low energies and a softening of the spring constant gradually until the saddle region, after which the levels become more spaced going to the limit mentioned above. Thus one can see that the procedure indicated describe the main characteristics of the energy spectrum of the system.

3. Summary of the Lipkin-Meshkov-Glick model

The LMG model was introduced in nuclear physics because it is simple enough to be solved exactly but is yet non-trivial [8] and it has been used to test approximation methods of many body systems [9]. Also, it has played an important role in other fields of physics like, for example, in condensed matter studies to study the size scaling for infinitely coordinate systems [10, 11], in the description of a two-mode Bose-Einstein condensates [12], and in quantum optics by Kitagawa and Ueda [13] to produce spin squeezed states. In the context of condensed matter describing mutually interacting 1/2 spins embedded in a magnetic field, entanglement entropy and phase transitions in collective spin systems, in particular in the LMG model has attracted much attention. This quantum information concept (entanglement entropy) displays a singularity at the critical point as a function of the magnetic field, the anisotropy parameter, and the size of the system [14]. Using the Holstein-Primakoff representation and the continuous unitary transformation technique finite size scaling exponents for several observables of the LMG model were calculated [15, 16, 17, 18]. Entanglement properties of the ground state and its dynamics have been analyzed under different approaches, remarking important properties of the system [19, 20, 21, 22, 23].

The model Hamiltonian is given by [8]

$$H = \epsilon J_0 + \frac{\lambda}{2}(J_+^2 + J_-^2) + \frac{\gamma}{2}(J_+ J_- + J_- J_+) \ , \tag{8}$$

where the λ-term annihilates pairs of particles in one level and creates pairs in the other level while the γ-term scatters one particle up and another one down. The Hamiltonian can be rewritten in the form

$$H = \epsilon J_0 + \bar{\gamma}_x \, J_x^2 + \bar{\gamma}_y \, J_y^2 \ , \tag{9}$$

where $\bar{\gamma}_x = \gamma + \lambda$ and $\bar{\gamma}_y = \gamma - \lambda$.

As it is usual, for simplicity only the case with maximum symmetry $J = N/2$, with N the total number of particles is considered. Exact eigenvalues and eigenvectors of the Hamiltonian (9), can be obtained in the basis $\{|N\,n\rangle\}$, with n varying from 0 to N, defined as

$$|N\,n\rangle = \sqrt{\frac{(N-n)!}{N!\,n!}}J_+^n|N\,0\rangle\ . \tag{10}$$

The state $|N\,0\rangle$ is the unperturbed ground state, i.e. the ground state when $\lambda = \gamma = 0$ because we are considering the parameter $\epsilon > 0$. In this case all the particles occupy the lowest single particle state. Since the Hamiltonian (9) has non-vanishing matrix elements between states (10) with $n \to n, n \pm 2$ its eigenfunctions can be written in terms of linear combinations of states with even (e) or odd (o) values of n (e.g. the number of particle-hole pairs), namely:

$$|\alpha, e\rangle = \sum_{i=0}^{N} C^{\alpha}_{i,e}|N\,2\,i\rangle\ , \qquad |\alpha, o\rangle = \sum_{i=0}^{N-1} C^{\alpha}_{i,o}|N\,2i+1\rangle\ . \tag{11}$$

In this notation α is the eigenvalue-index, thus $|\alpha = 1, e\rangle$ and $|\alpha = 1, o\rangle$ are the exact ground state and the exact first excited state, respectively.

Following the method mentioned in the previous section, in [24, 25] the separatrix of the system was found taking the expectation value of the Hamiltonian with respect to the SU(2) coherent states, i.e.,

$$\mathcal{H} = \langle\zeta|H|\zeta\rangle\,,$$

where $|\zeta\rangle = (1 + |\zeta|^2)^{-J}\exp(\zeta J_+)|0\rangle$, and its stability behavior was determined through the catastrophe formalism. $\mathcal{H}$ represents the classical Hamiltonian function of the model in the limit when $N \to \infty$. A detailed analysis of the classical and quantum dynamics of the LMG model can be found in [25].

The separatrix of the LMG model is given in Fig. 3, and the behavior of the corresponding eigenstates is also indicated with the names collective (bimodal) and independent particle (unimodal). In the independent particle regime the ground state is unique, while in the collective one has degeneracy two. The arrows indicate trajectories in the control parameter space which cross quantum phase transitions, they are of second order when the separatrix is crossed from the collective regime to the single particle one and of first order for the crossing between two collective ones. Another fact related with the LMG model is that the crossing of the point $(\gamma_x = -1, \gamma_y = -1)$, which is not shown in Fig. 3, along the straight line $\gamma_y = -\gamma_x - 2$ constitutes a third order phase transition. In that point a convergence of second phase transitions are happening, which it is currently under investigation.

It is important to mention that trial states with good parity yield a better description of the ground and first excited states of the LMG model. These trial states are constructed as Schrödinger cat states in terms of combinations of spin coherent states [25]. The calculation of even and odd energy surfaces were shown to closely reproduce the exact quantum behavior of the ground and first excited states, and their crossings. The intersection of these even and odd energy surfaces yields another Maxwell set $\gamma_{y0} = (1/\gamma_x)$; in the region between the branches of the hyperbola there are no crossings between energy levels, whereas in the exterior part degeneracy is present.

4. Fidelity

In classical information theory emerges the concept of *fidelity*. It is a basic ingredient of communication theory, which measures the accuracy of a transmission. For example, one can transmit a sequence of N signals and it is important to know the probability, over all possible

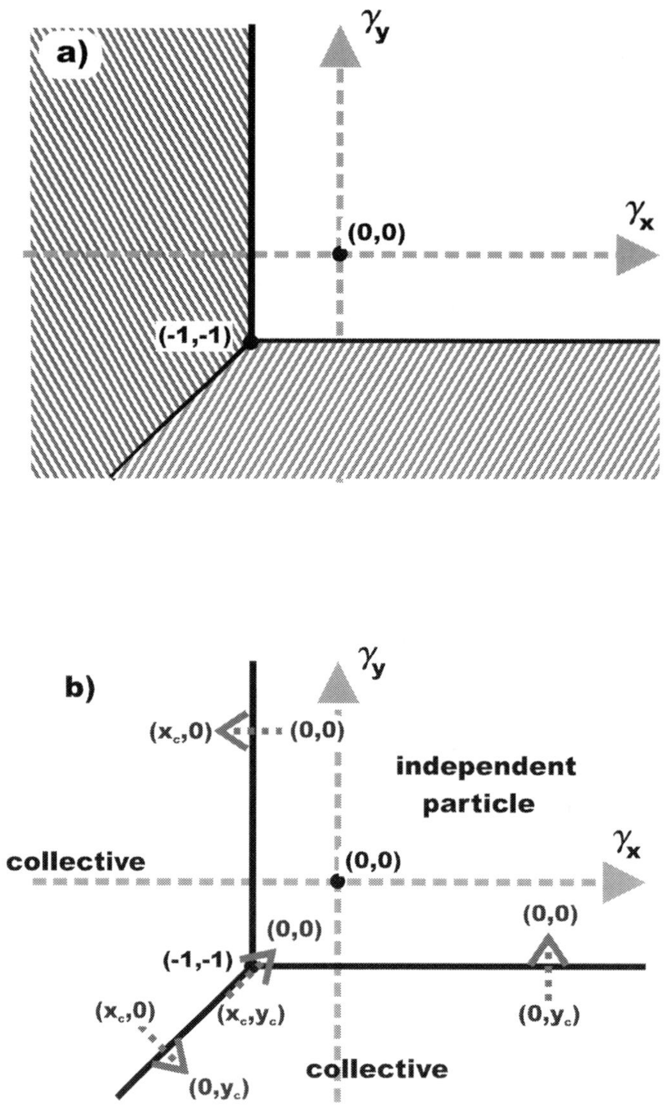

Figure 2. a) Bifurcation sets are shown with black continuos lines. The Maxwell sets are displayed filled up with negative slope (red) lines for equally deep minima along the axes x, $(\pm x_c, 0)$, and with positive slope (blue) lines for minima along the y axis, $(0, \pm y_c)$. b) The shape phase diagram for the LMG model is established and the quantum phase transitions indicated. There are second order transitions for $(0,0) \leftrightarrow (x_c, 0)$, $(0, y_c) \leftrightarrow (0,0)$, and $(x_c, y_c) \leftrightarrow (0,0)$; and first order transitions for the case $(x_c, 0) \leftrightarrow (0, y_c)$. Notice that $\gamma_x \equiv \frac{2J-1}{\epsilon}\bar{\gamma}_x$ and $\gamma_y \equiv \frac{2J-1}{\epsilon}\bar{\gamma}_y$.

Journal of Physics: Conference Series **387** (2012) 012021 doi:10.1088/1742-6596/387/1/012021

N-sequences, that the received signals coincide with the sent one; this is the definition of the fidelity. This concept is also closely related with Shannon's noiseless coding theorem and channel capacity, which play a fundamental role in classical information theory. The first one answers the question: how much can a message be compressed while still obtaining the same information?, being the answer the calculation of the corresponding Shannon's entropy [3]. In 1994, a similar theorem was developed by Schumacher for quantum communication as an extension of Shannon's noiseless coding theorem [26], where the von Neumann entropy plays the role of the Shannon entropy for the classical case.

For two pure states described by the density matrices $\rho_1 = |\Psi_1\rangle\langle\Psi_1|$ and $\rho_2 = |\Psi_2\rangle\langle\Psi_2|$ the fidelity between them is defined as follows [27]

$$F = |\langle\Psi_1|\Psi_2\rangle|^2 . \tag{12}$$

In quantum physics, an overlap between two quantum states denotes the transition amplitude from one state to other, while in the information formalism it measures the similarity between two states. Then the overlap between the input and output states becomes a useful measure of the loss of information during the transportation of a message.

Recently [28], the fidelity has been proposed as a tool to determine the quantum phase transitions of a physical system, the motivation being related directly with the meaning of a quantum phase transition. A phase transition in the control parameter space takes place when the ground state of the system changes significantly when one or several of the control parameters are varied. Therefore, one can fix all the control parameters except one, and the quantum phase transition will be localized at point where

$$F(\gamma, \gamma + \delta\gamma) = |\langle\Psi(\gamma)|\Psi(\gamma + \delta\gamma)\rangle|^2 \tag{13}$$

reaches its minimum value.

In this work, we generalize the definition of the fidelity by considering the overlap between the exact quantum state and an arbitrary $|\Phi\rangle$ representative state of the Hilbert space of the physical system,

$$F(\Phi, \Psi(\gamma)) = |\langle\Phi|\Psi(\gamma)\rangle|^2 . \tag{14}$$

By means of this calculation we are able to determine the position in the varying control parameter related with the quantum phase transitions due to crossings or anticrossings of the energy levels in a quantum system. As an example of the procedure, we are going to show this for the LMG model.

First of all, we consider the diagonal case of the LMG model, that is $\gamma_x = \gamma_y$. It has been shown that there are crossings of the energy spectra for the values [25]

$$\gamma_k^{(1)} = -\frac{2J-1}{2J-1-2k}, \qquad \gamma_k^{(2)} = -\frac{2J-1}{2J-2-2l}, \tag{15}$$

with $k = 0, 1, \cdots 2J-1$ and $l = 0, 1, \cdots 2J-2$. Then, if for example we consider $N = 20$ particles distributed into the two-level system one has two sets of crossings, the first one associated to the crossings between the ground and first excited states, and the second one between the ground and second excited states,

$$\begin{aligned} S_1 &= \left\{-1, -\frac{19}{17}, -\frac{19}{15}, -\frac{19}{13}, -\frac{19}{11}, -\frac{19}{9}, -\frac{19}{7}, -\frac{19}{5}, -\frac{19}{3}, -19\right\}, \\ S_2 &= \left\{-\frac{19}{18}, -\frac{19}{16}, -\frac{19}{14}, -\frac{19}{12}, -\frac{19}{10}, -\frac{19}{8}, -\frac{19}{6}, -\frac{19}{4}, -\frac{19}{2}\right\}, \end{aligned}$$

Journal of Physics: Conference Series **387** (2012) 012021 doi:10.1088/1742-6596/387/1/012021

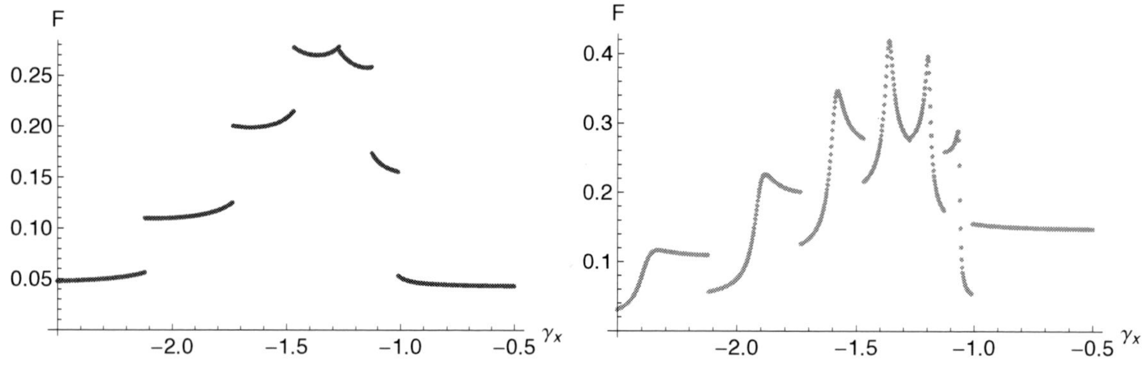

Figure 3. At the left, the fidelity between the even ground state of the LMG model with a spin coherent state with $\zeta = 0.9$ is shown. At the right, the fidelity between the odd ground state of the model with the same spin coherent state is displayed.

and their corresponding positive values because the model Hamiltonian is symmetric under the interchange of sign of the control parameter γ_x.

Fig. 3 shows six sudden changes in the values of the fidelity between a spin coherent state, with $\zeta = 0.9$, and the ground state of the even parity case. We propose that these changes are defining quantum phase transitions in the LMG model for the diagonal case. Comparing these with the known crossings above, one finds that there is an exact agreement with the crossings indicated by the set S_1, in the region $-2.5 \leq \gamma_x \leq -0.5$. For the overlap between the same spin coherent state with the odd eigenstate of the LMG model one finds twelve sudden changes, and again one has a complete agreement with the crossings between the energy levels indicated in the sets S_1 and S_2.

For the non-diagonal case, we evaluate the fidelity between the (trivial) test functions for $N = 20$ particles of the form

$$\begin{aligned} |\phi_T\rangle_{even} &= \frac{1}{1540}\,(0,2,4,6,8,10,12,14,16,18,20)\,, \\ |\phi_T\rangle_{odd} &= \frac{1}{1330}\,(1,3,5,7,9,11,13,15,17,19)\,. \end{aligned}$$

with respect to the ground and first excited states of the even and odd parity cases. We selected $\gamma_y = \gamma_x + 0.015$, and the fidelities between the even and odd first excited states with respect to the corresponding test functions are exhibited in Fig. 4. From the results, one notices immediately the localization of four sudden changes in the value of the fidelity for the even ground state while for the first excited state there are seven sudden changes in the fidelity. One can also see that the even first excited state detects the quantum phase transitions present in the even and odd ground states of the LMG model. Therefore, the calculation of the fidelity is able to detect all the phase transitions present in the LMG model for the diagonal and non-diagonal cases.

We have shown that the fidelity is a good tool to determine the localization of the quantum phase transitions, when a reasonable test function is used to calculate the fidelity of the quantum ground state with respect to it, and determine the localization of the quantum phase transitions.

It is also interesting to calculate the fidelity in the form $F(\gamma, \gamma + \delta\gamma)$ and look for the minima. This implies the determination of the second derivative of the fidelity as follows

$$\chi_F = 2\frac{1 - F(\gamma, \gamma + \delta\gamma)}{\delta\gamma^2}\,, \tag{16}$$

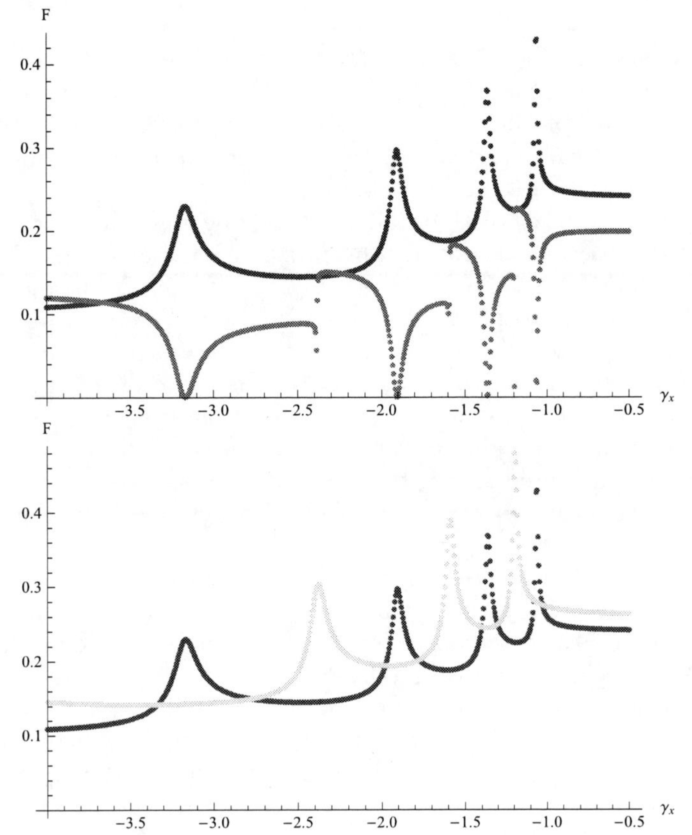

Figure 4. At the top, the fidelity between the even ground state, upper (blue) line, and even first excited state, lower (red) line, of the LMG model with respect to the $|\phi_T\rangle_{even}$ test function is shown. At the bottom, the fidelity between the even parity, lower (blue) lines, and odd parity, upper (cyan) line, ground states of the model with respect to the $|\phi_T\rangle_{odd}$ test function is displayed.

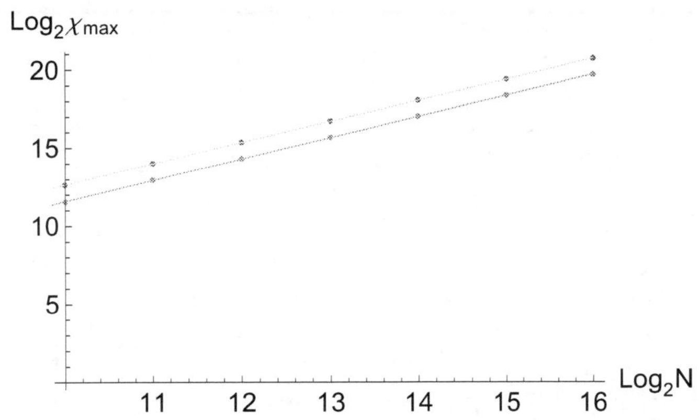

Figure 5. We display the plots $\text{Log}_2\chi^e_{Fmax}$ vs $\text{Log}_2 N$, upper (green) line and $\text{Log}_2\chi^o_{Fmax}$ vs $\text{Log}_2 N$, lower (orange) line, for $\gamma_y = 0.5$. The lines indicate the corresponding best fits, and one can see that they have approximately the same slope.

which expresses the answer of the physical system to the interaction term associated to the control parameter γ, and for that reason it has been called the fidelity susceptibility. The maximum of the fidelity susceptibility determines the exact localization of the quantum phase transition together with critical exponents of the quantum phase transition associated to the corresponding interaction term of the Hamiltonian [29].

This new concept can be used to determine the scaling behavior and universality of the quantum phase transition [4, 29]. Following the procedure indicated in [15] the LMG model Hamiltonian can be expanded in terms of powers of N^{-1}, $N^{1/2}$ and N^0 by means of the Holstein-Primakoff representation of the angular momentum generators and determine exact expressions for the fidelity susceptibility in both sides of the quantum phase transition. These results let us determine the scaling properties and universality of the quantum phase transitions. In the case of the first order phase transitions the fidelity is zero and then the procedure cannot be applied while for the second order quantum phase transitions it works perfectly. To show this, we study the LMG model for $\gamma_y = -0.5$ and change the parameter γ_x associated to the component J_x^2 of the Hamiltonian in a range close to the quantum phase transition determined previously with an arbitrary test function as it was mentioned above. In this way one is able to establish the range of variation for the control parameter to $-1.03 \leq \gamma_x \leq 0.998$.

As we are interested in finding the scaling properties of the quantum phase transition associated to the interaction Hamiltonian J_x^2, the LMG model Hamiltonian is diagonalized for a number of particles 2^N with $N = \{10, 11, 12, 13, 14, 15, 16\}$. Then we calculate the fidelity susceptibility for the even and odd parity cases, by means of the expression (16), for $\gamma_y = 0.5$ and $-1.03 \leq \gamma_x \leq -0.998$. The results are displayed in Fig. 5, where the best fits of the corresponding even and odd fidelity susceptibilities are indicated by the upper (green) and lower (orange) lines, respectively. For the even case the slope is given by 1.348 and the ordinate to the origin by -0.836, while for the odd case one gets the slope 1.356 and the ordinate -1.997. Notice that for a small number of particles of the order of 2^8 and smaller there are deviations of this universal rule.

5. Conclusions

We have made a brief description of phase transitions in classical mechanics, and how it can be extended to characterize the quantum phase transitions by means of the catastrophe formalism together with the spin coherent states. We have shown that the quantum information concepts as the fidelity and fidelity susceptibility are very useful to determine the quantum phase transitions present in the LMG model together with universal properties of for a very large number of particles. We concluded also that by means of almost *any* test function, the fidelity can be used to determine the approximate values of the control parameters that yield quantum phase transitions for a finite number of particles. This lets us determine the position in the control parameter space of the crossings and anti-crossings present in the LMG model, for a finite number of particles. Thus, it will be important to study the behavior of the different physical observables of the LMG model at these finite quantum phase transitions.

The scaling behavior of a quantum transition can be estimated for second order quantum phase transitions using the fidelity susceptibility concept. This cannot be done for first order quantum phase transitions because no matter how small is the difference $\delta\gamma$ the states are orthogonal. This is known in the literature of many-body physics as the Anderson orthogonality catastrophe [30]. The scaling properties of the quantum phase transitions are shown in Fig. 5 for the parameter value $\gamma_y = 0.5$; analogous results have been found for $\gamma_y = -0.5$, and in the vicinity of the phase transition $\gamma_x = -1$.

The specific susceptibility defined as

$$\chi_{spe} = \frac{\chi_{Fmax}}{\chi_F(\gamma_x)} - 1 , \tag{17}$$

Journal of Physics: Conference Series **387** (2012) 012021 doi:10.1088/1742-6596/387/1/012021

can be plotted in terms of the product of the control parameter $(\gamma_x - \gamma_x^c)N^{2/3}$. It exhibits a universal behavior for any large number of particles, as it was shown in [29], although in our case a different interaction term of the LMG model Hamiltonian was considered.

Acknowledgments

This work was partially supported by CONACyT-México, with reference number 101541, and by DGAPA-UNAM under project IN102811.

References

[1] Gor'kov L P 1959 *Soviet Physics JETP* **9** 1364
[2] Wilson K G 1975 *Rev Mod Phys* **47** 773
[3] Benenti G, Casati G, and Strini G 2007 *Principles of quantum computation and information* Vol II (World Scientific, Singapore)
[4] Gu S H 2010 *Int J Mod Phys B* **24** 4371
[5] Arnol'd V I 1992 *Catastrophe Theory* (Springer-Verlag)
[6] Gilmore R 1981 *Catastrophe Theory for Scientists and Engineers* (Wiley, New York)
[7] Gilmore R, Kais S, and Levine R 1986 *Phys Rev A* **42** 2442
[8] Lipkin H J, Meshkov N, and Glick A J 1965 *Nucl Phys* **62** 188, 199, 211
[9] Hirsch J G, Civitarese O, and Reboiro M 1999 *Phys Rev C* **60** 24309
[10] Botet R, Jullien R, and Pfeuty R 1982 *Phys Rev Lett* **49** 478
[11] Botet R and Jullien R 1983 *Phys Rev B* **28** 3955
[12] Pérez C, González-Alonso J R, Castaños O, and López-Peña R 2010 *Ann Phys* **325** 325
[13] Kitagawa M and Ueda M 1993 *Phys Rev A* **47** 5138
[14] Latorre J I, Orus R, Rico E, and Vidal J 2005 *Phys Rev A* **71** 064101
[15] Dusuel S and Vidal J 2004 *Phys Rev Lett* **93** 237204; 2005 *Phys Rev B* **71** 224420
[16] Dusuel S, Vidal J, Arias J M, Dukelsky J, and García-Ramos J E 2005 *Phys Rev C* **72** 064332
[17] Leyvraz F and Heiss W D 2005 *Phys Rev Lett* **95** 050402
[18] Heiss W D, Scholtz F G, and Geyer H B 2005 *J Phys A: Math Gen* **38** 1843
[19] Wu L-A, Sarandy M S, and Lidar D A 2004 *Phys Rev Lett* **93** 250404
[20] Wu L-A, Sarandy M S, Lidar D A and Sham L J 2006 *Phys Rev A* **74** 052335
[21] Vidal J, Palacios G and Aslangul C 2004 *Phys Rev A* **70** 062304
[22] Menotti C, Anglin J R, Cirac J I and Zoller P 2001 *Phys Rev A* **63** 023601
[23] Micheli A, Jaksch D, Cirac J I and Zoller P 2003 *Phys Rev A* **67** 013607
[24] Castaños O, López-Moreno E and López-Peña R 2003 *Rev Mex Fís.* **49** S4, 15
[25] Castaños O, López-Peña R, Hirsch J G and López-Moreno E 2005 *Phys Rev B* **72** 012406; 2006 *Phys Rev B* **74** 104118
[26] Schumacher B 1995 *Phys Rev A* **51** 2738
[27] Jozsa R 1994 *J Mod Optics* **41** 2315
[28] Zanardi P and Paunkovic N 2006 *Phys Rev E* **74** 031123
[29] Kwok H, Ning W, Gu S and Lin H 2008 *Phys Rev E* **78** 032103
[30] Anderson P W 1967 *Phys Rev Lett* **18** 1049

AUTHOR INDEX

Aguilera, E. 1, 9
Aubin, S. 24
Avogadro, P. 111
Badagnani, D. 73
Bahri, C. 127
Ball, G. 103
Barbero, C. 73
Batchelder, J. 29
Behr, J. 24
Berg, G. 17
Bijker, R. 79
Brewer, N. 29
Camacho, A. 9
Canales, F. 53
Castanos, O. 157
Civitarese, O. 48
Collister, R. 24
Draayer, J. 121, 127
Dreyfuss, A. 127
Dytrych, T. 121, 127
Ebata, S. 111
Ferretti, J. 79
Fujita, Y. 17
Galindo-Uribarri, A. 95
Gomez, E. 24
Gorres, J. 17
Gross, C. 29
Grzwacz, R. 29
Guo, L. 111
Gwinner, G. 24
Hamilton, J. 29
Harakeh, M. 17
Hardy, J. 39
Hatanaka, K. 17
Hess, P. 141
Hirsch, J. 147, 157
Inakura, T. 111
Karny, M. 29
Kuzniak, A. 29
Langr, D. 121, 127
Launey, K. 121, 127
Lei, Y. 132
Liu, S. 29
Long, A. 17
Lopez-Pena, R. 157
Madurga, M. 29
Mariano, A. 73
Mendez, A. 29
Miernik, K. 29
Miller, D. 29
Mondragon, A. 53
Morales-Hernandez, G. 141
Nahmad-Achar, E. 157
Nakatsukasa, T. 111
Neveling, R. 17
Orozco, L. 24
Padgett, S. 29
Paulauskas, S. 29
Pearson, M. 24
Pittel, S. 132
Poves, A. 132
Ramayya, A. 29
Rykaczewski, K. 29
Sandulescu, N. 132
Santopinto, E. 79
Sheng, D. 24
Smit, F. 17
Srivastava, P. 147
Stracener, D. 29
Stuchbery, A. 86
Suhonen, J. 48
Talwar, R. 17
Tamil, A. 17
Tandecki, M. 24
Thakur, B. 132
Towner, I. 39
Wlescher, M. 17
Wolinska-Cichocka, M. 29
Wong, C. 63
Yepez-Martinez, H. 141
Yoshida, K. 111
Zganjar, E. 29
Zhang, J. 24
Zhao, Y. 132